The Progressive Environmental Prometheans

William B. Meyer

The Progressive Environmental Prometheans

Left-Wing Heralds of a "Good Anthropocene"

palgrave
macmillan

William B. Meyer
Hamilton, New York, USA

ISBN 978-3-319-29262-5 ISBN 978-3-319-29263-2 (eBook)
DOI 10.1007/978-3-319-29263-2

Library of Congress Control Number: 2016945985

Printed on acid-free paper

This Palgrave Macmillan imprint is published by Springer Nature
The registered company is Springer International Publishing AG Switzerland

In memory of Andrei V. Bell (1953–2014)
Prosti menia, liubov' moei zhizni!

Acknowledgements

I am indebted to Colgate University for a faculty leave that helped me complete this book and to comments and questions by attendees at a Social Sciences Division colloquium in November, 2013.

Contents

1

Introduction: Earthviews and Worldviews

If we found environmentalists and their opponents scattered randomly along the conventional spectrum of political ideologies running from left to right, there would be little point in asking how the two realms of beliefs—call them earthviews and worldviews—are related. But the question does arise, for support for green causes and measures today is far from being randomly distributed; it clusters on the left and resistance to it on the right. The American members of Congress rated most highly by environmental organizations enjoy equally high ratings from liberal watchdog groups and low ones from their conservative counterparts. Green parties in Europe have gained most of their support, formed most of their alliances, and staked out most of their positions on the left; they have incurred the most animosity from the right.[1] The rule has its

[1] For Europe, see Jean Jacob, *Histoire de l'écologie politique* (Paris: Albin Michel, 1999); Neil Carter, *The Politics of the Environment: Ideas, Activism, Policy* (Cambridge: Cambridge University Press, 2001), 72–76; Jon Burchell, *The Evolution of Green Politics: Development and Change within European Green Parties* (London: Earthscan, 2002); Herbert Kitschelt with Anthony J. McGann, *The Radical Right in Europe: A Comparative Analysis* (Ann Arbor, MI: University of Michigan Press, 1995), 22 (noting "the frequent vilification of feminist and environmentalist movements" by the extreme right); and Kostas Gemenis, Alexia Katsanidou, and Sofia Vasilopoulou, "The Politics of

W.B. Meyer, *The Progressive Environmental Prometheans*,
DOI 10.1007/978-3-319-29263-2_1

exceptions and anomalies, to be sure, which include a certain number of avowed conservative environmentalists, Marxist, and populist critiques of a neo-Malthusianism focused on the dangers of population growth, and challenges to some of the priorities of mainstream green organizations by the environmental justice movement.[2] It holds generally true nonetheless. "Environmental and natural-resource doomsayers," the conservative American jurist Richard Posner wrote in 2001, "are invariably leftists." "Ever since the environment was put on the global agenda some four decades ago," two Canadian political scientists have summed up the matter, "the left has attributed more importance to this issue than the right … the main cleavage places environmentalists on the left and their opponents on the right."[3]

Anti-Environmentalism: Positional Framing by the European Radical Right," paper prepared for the MPSA Annual Conference, Chicago, 2012. For 2013, the most recent full year for which ratings were available at the time of writing, twenty-seven members of the US Senate received perfect scores of 100 from the League of Conservation Voters (all of them Democrats except for the Senate's only independent, and only avowed socialist, Bernard Sanders of Vermont). Their median rating from Americans for Democratic Action was ninety-five and their mean ninety-four; their median American Conservative Union rating was four, as was their mean. The five senators (all Republicans) who received LCV scores of zero enjoyed an ADA median of zero and mean of one, and an ACU median of ninety-two and mean of ninety-three. Data from scorecard.lcv.org/, acuratings.conservatve.org/acu-ratings-chart/ and adaction.org/pages/publications/voting-records.php.

[2] For conservative environmentalism, see, for example, John Gray, "An Agenda for Conservative Environmentalism," in *Beyond the New Right: Markets, Government and the Common Environment* (London: Routledge, 1993), 124–177; John R. E. Bliese, *The Greening of Conservative America* (Boulder, CO: Westview Press, 2001); and Roger Scruton, *How to Think Seriously About the Planet: The Case for an Environmental Conservatism* (New York, NY: Oxford University Press, 2012). For leftist critiques of neo-Malthusianism, see, for example, David Harvey, "Population, Resources, and the Ideology of Science," *Economic Geography* 50, #3 (1974), 256–277 and *Spaces of Capital: Towards a Critical Geography* (Edinburgh: Edinburgh University Press, 2001) and Hans Magnus Enzensberger, "A Critique of Political Ecology," in Ted Benton, ed., *The Greening of Marxism* (New York, NY: Guilford Press, 1996), 17–49. (For the earlier Marxist rejection of Malthus, see Ronald L. Meek, ed., *Marx and Engels on Malthus* (New York, NY: International Press, 1954).) For the environmental justice critique, see, for example, Eileen McGurty, *Transforming Environmentalism: Warren County, PCBs, and the Origins of Environmental Justice* (New Brunswick, NJ: Rutgers University Press, 2007) and Ronald Sandler and Phaedra C. Pezzullo, eds., *Environmental Justice and Environmentalism: The Social Justice Challenge to the Environmental Movement* (Cambridge, MA: MIT Press, 2007).

[3] Richard Posner, *Public Intellectuals: A Study of Decline* (Cambridge, MA: Harvard University Press, 2001), 283; Alain Noël and Jean-Philippe Thérien, *Left and Right in Global Politics* (New York, NY: Cambridge University Press, 2008), 210. Noël and Thérien likewise cited Robert Paehlke's assertion, in *Environmentalism and the Future of Progressive Politics* (New Haven, CT: Yale University Press, 1989), 309n1, that although environmentalism could, in principle, be combined with any political ideology, "in fact it has most often been linked with the moderate left."

No one, I suppose, would argue that the pattern exists by mere chance. There must, then, be reasons for it. Evidently, something about the political ideologies of the contemporary left and right makes the former more sympathetic to environmental concerns than the latter. That such a difference exists may seem to be cause for dismay. Why, it might be asked, should one's beliefs about politics or society affect what—or whom—one believes about the relation of the physical, chemical, and biological processes of the earth's surface to the survival and well-being of its human population? How could they do anything but confuse and distort it? But in practice, however things ought to be in principle, they appear not to be irrelevant after all. In some way or another, most people's environmental beliefs are not merely scientific but political. Why they display the politics that they do is worth trying to clarify.

One can approach this question in several ways. One, with which I begin, is to inquire into the logic of today's divide and ask why concern about human impact on the earth should be so much greater on the left than on the right. Another, to which I devote the rest of the book, is to see whether the same alignments existed in earlier times as well. Answering the second question may help in arriving at or verifying an answer to the first. If there is some necessary and inherent logic to today's pattern, one would expect to find it stable over the course of history. If one did not find it so, substantially different explanations would be in order.

Some key terms require definition at the outset. First of all, what is environmentalism? I use the word to denote a commitment to several core beliefs: the irreplaceability of the natural world, in its basic contours, as a human home and the danger that excessive human pressures will disrupt its essential functions. The French ecologist and green activist Antoine Waechter spoke of the need for humankind "to place limits on its own expansion and on its own power" rather than suicidally seek to "push ever farther the bounds of human mastery." The American ecologist and green activist Barry Commoner wrote that the "ecosphere," "the home that life has built for itself on the planet's outer surface," "sustains people and everything that they do," but that human activities disharmonious with its workings have left it "so heavily strained that its continued stability is threatened" and it "is being driven towards collapse." More succinctly still, he proposed the axiom "nature knows best" as a basic statement of the essential green position. Aldo Leopold likewise proposed that an act is good if it helps to maintain, and bad if it undermines, the existing ecosystem: the assemblages of plant and animal life and the conditions of climate, water, terrain, soil, and material and energy flow on which they—and ultimately we—depend. In the formulation of the French green philosopher Michel Serres, "we have become so much and so little masters of the Earth that it once again threatens to master us in turn," human dominance of the environment being a mirage and a dangerous one. Serres proposed a "natural contract" by analogy to the "social contract," replacing human efforts to subordinate the earth with a relation of harmonious coexistence as essential to society's own survival.[4]

On this view, the structure and processes of the environment as it is and as it has been possess a logic that may not always be evident but that should always be presumed. Human action that substantially alters them, whether on purpose or not, therefore carries a strong potential for harm, not least to human beings themselves. Environmentalists see the biophysical world as an immensely complex network of complex ecosystems, one whose present state reflects a long process of development through their internal and mutual adjustment. Any component of the environment may play key roles, some of them far from obvious, in maintaining the narrow range of conditions that make the earth's surface a livable habitat for human beings. Therefore, further alterations in the environment or further increases in society's demands (enormous already in their magnitude and scope) are apt to disrupt it, and the more dangerously the more novel, sudden, drastic, profound, and far-reaching they are. What I call environmentalism corresponds closely to what the American political scientist Paul Wapner has described as "naturalism" or "the dream of naturalism," which, in his words, "recommends that we align ourselves with, rather than impose ourselves on, the natural world"—and as much for our own good as for its.[5]

---

[4] Antoine Waechter, *Dessine-moi une planète: L'écologie, maintenant ou jamais* (Paris: Albin Michel, 1990), 156–157, 222; Barry Commoner, *The Closing Circle: Nature, Man, and Technology* (New York, NY: Knopf, 1971), 11–12, 41–45, 112; Aldo Leopold, *A Sand County Almanac & Other Writings on Ecology and Conservation*, ed. Curt Meine (New York, NY: Library of America, 2013), p. 188; Michel Serres, *The Natural Contract*, trans. Elizabeth MacArthur and William Paulson (Ann Arbor, MI: University of Michigan Press, 1995), 33.

[5] Paul Wapner, *Living Through the End of Nature: The Future of American Environmentalism* (Cambridge, MA: MIT Press, 2010) (quotation from 75; see also his Chap. 3, "The Dream of

A statement like Commoner's "nature knows best" is an extreme one, and meant to provoke. No one follows it as if it were absolutely and literally true. And as many writers have emphasized, environmentalists have often exaggerated the balance and equilibrium of earth systems even when largely undisturbed by human actions and ignored the fact that there is no single stable natural condition that we can use as a standard and seek to maintain or return to.[6] Those who assert nature's wisdom are, rather, issuing admonitions to stop and think before acting. They are warning that there is much about the environment that even scientific experts do not know—and, in a phrasing that has lately come into currency, cannot ever fully know—save only for a few general truths: that it is complex in its workings, though finite in its bounty, and that we cannot survive as a species if we change it too much. To suppose that "people

---

Naturalism," 53–76). More concretely, a group of French ecologists in the 1970s chose Diogenes to symbolize the environmentalist virtues of living simply and in accordance with nature: Jacob, *Histoire de l'écologie politique*, 128–130.

[6] For important early statements of this point, see Daniel Botkin, *Discordant Harmonies: A New Ecology for the Twenty-First Century* (New York, NY: Oxford University Press, 1990) and Alston Chase, *In a Dark Wood: The Fight Over Forests and the Tyranny of Ecology* (Boston, MA: Houghton Mifflin, 1995). Much of the critique is valid, but it does not merely make environmentalism (nor, for that matter, Prometheanism) as I have defined it an erroneous or confused way of thinking, for several reasons. First, without exception, the figures whom the main chapters of this book focus on lived before the critique was articulated, and could thus speak unself-consciously of the natural versus the human-altered state of the earth, without having to face the question of which of multiple natural states they meant or of what elements of human impact they already incorporated. Second, the critique, if accepted, does not mean that all possible human-altered ones are equally distant from those of any period that were not human-altered, or that all human-altered states are equally compatible with the long-term survival of the remaining elements that have not been drastically altered. Even if, today, no feature of the earth's surface is "natural" in the sense of being entirely unaffected by human action, differences of degree exist; some are still more so than others. And as one proponent of the "new ecology" observes, "human-induced ecological disturbances … differ from natural ones in frequency, magnitude, and depth": Karl S. Zimmerer, "Human Geography and the 'New Ecology': The Prospect and Promise of Integration," *Annals of the Association of American Geographers* 84, #1 (1994), 116. The greater "violence, rapidity, and scope" of human-induced than of non-human-induced change still represent, as they did for Aldo Leopold (quoted by Bryan G. Norton, *Toward Unity among Environmentalists* (New York, NY: Oxford University Press, 1991, 52), criteria for concern about it that do not presuppose some overarching balance or equilibrium in nature. (See also Donald Worster, "The Ecology of Order and Chaos," *Environmental History Review*, 14, #1/2, 1990, 1–18.) Nor, finally, do environmentalists have to rely on the empirically and metaphysically questionable standard of the "natural" state. Possible alternatives include measures of "ecological integrity": David Pimentel, Laura Westra, and Reed F. Noss, eds., *Ecological Integrity: Integrating Environment, Conservation, and Health* (Washington, DC: Island Press, 2000).

know best," they imply, is dangerous hubris too easily indulged in, for many changes that seem obviously improvements may well turn out to be the opposite. We should, therefore, restrain and limit our impacts, deliberate as well as inadvertent, as much as we can. What some now call "the end of nature" and others the advent of the "Anthropocene" does not fatally undermine the environmentalist credo. That human action has now profoundly altered much of the face and functioning of the planet, leaving little that can be called "nature" in the strictest senses of the word, is not, to environmentalists, any reason why still further and deeper change from the present state of the earth should be accepted. Indeed, it is a strong reason why such change is more than ever to be feared.[7] That rational management and domestication of the earth can make the era of human domination the "good Anthropocene" that some writers now envision, environmentalists regard as a dangerous delusion.

Thus one of the major themes of modern environmentalism is that of the unexpected and usually unwelcome consequences of human actions. (I am speaking here, and from now on, of anthropocentric environmentalism, or that form that frankly aims to secure the survival and well-being of human beings and of elements of their surroundings that they value; ecocentric forms of environmentalism add the rights of non-human entities.) In the language of systems theory, the environment is pervaded by feedback loops. In meddling with it at one point, we may unwittingly set some of them in motion and produce results quite unlike those that we expected. Simple prudence, then, should make us limit the extent, depth, and speed of our interference as much as we reasonably can and undertake novel kinds of changes only with great reluctance. By no accident, "ecology," strictly speaking the name of a branch of natural science, has in ordinary usage become a synonym for environmentalism, which sees the interactions and complexities that it studies as crucially important. The Great Books of modern environmentalism rest much of their case on the unintended-consequences argument. Rachel Carson's *Silent Spring* (1962) painted a horrific picture of the side effects on non-targeted spe-

[7] See, for example, David Ehrenfeld, "The Fable of Managed Earth," 85–108 and George Wuerthner, "Why the Working Landscape Isn't Working," 162–173, in George Wuerthner, Elieen Crist, and Tom Butler, eds., *Keeping the Wild: Against the Domestication of Earth* (Washington, DC: Island Press, 2014).

cies, people among them, produced by the unfettered spraying of long-lived synthetic chemical pesticides. Those who used so powerful a tool so extensively for the single purpose of destroying unwanted insects, Carson wrote, ignored the reality that the natural world was one "of ecology, of interrelationships, of interdependence." "We are accustomed to look for the gross and immediate effect and to ignore all else," she observed, but in such a complex world much else would always occur. She approvingly quoted a Dutch scientist to the effect that in our careless use of new technologies "we are walking in nature like an elephant in a china closet." Aldo Leopold warned to similar effect that "all land-use technologies … are encountering unexpected and baffling obstacles which show clearly that despite the superficial advances in technique, *we do not yet understand and cannot yet control* the long-time interrelations of animals, plants, and mother earth" (emphasis in the original).[7] The German green philosopher Hans Jonas decried a technological hubris that increasingly risked "setting the whole system of countless and delicate balances adrift toward catastrophe" by ignorantly disturbing ever more of them. "So long as we have not attained certainty of prediction," he continued, "and especially in view of the likely irreversibility of some of the initiated processes … *caution* is the better part of bravery."[8] Such reasoning underlies the "precautionary principle," which has been particularly influential in western Europe, calling for curbs on possibly harmful human actions even when proof that they will cause harm is not yet available.

A second and related theme of modern environmentalism is one of scale: the unsustainable magnitude of human demands on the earth's finite resources and the urgent need to restrain and eventually to reduce them. Thomas Robert Malthus remains an intellectual hero to many environmentalists for having emphasized, as early as 1798, the physical limits to the land's capacity for producing food.[9] His successors have extended his argument to other natural resources and to the "natural services" provided by the processes that maintain such background condi-

[8] Hans Jonas, *The Imperative of Responsibility: In Search of an Ethics for the Technological Age*, trans. Hans Jonas and David Herr (Chicago, IL: University of Chicago Press, 1984), 188, 191.

[9] On Malthus's stature among modern environmentalists, see Robert Mayhew, *Malthus: The Life and Legacies of an Untimely Prophet* (Cambridge, MA: Belknap Press of Harvard University Press, 2014).

tions as a stable climate and atmospheric composition necessary for life to continue. Humankind, the ecologists Paul and Anne Ehrlich warned, "is living largely on capital," the accumulated and exhaustible stocks of material, energy, and life-support that its actions are dissipating. The lesson they drew was the need "to reduce the scale of human activity."[10] The biosphere, neo-Malthusian environmentalists argue, the zone of the earth's surface within which alone human beings can live, is a system of a given size, capable of tolerating only certain levels of stress and alteration and subject to abrupt change—which would be catastrophic for civilization—if our demands exceed its limits. Its sustainable "carrying capacity" for human numbers and demands may be transgressed in the short run, but only at the price of disaster to follow. If Carson's book effectively dramatized the danger of unintended consequences, the collaborative study *The Limits to Growth* (1972) did as much, not least in its very title, to disseminate concerns about the finite capacity of the planet's systems to support consumption by rising human numbers and absorb and assimilate the wastes emitted by their activities. Modeling a number of possible pathways for world society under continued rapid growth, *The Limits to Growth* painted bleak pictures of overshoot and collapse, differing only in the immediate cause of catastrophe. Sometimes it was the exhaustion of nonrenewable resources, sometimes excessive pollution, sometimes the failure of agriculture to keep up with the demand for food. In the same year that *The Limits to Growth* saw print, a team of British ecologists published a similar manifesto entitled *A Blueprint for Survival*. Society's large and rising resource demands, pollution emissions, and conversion of land for direct human use, it warned, menaced humankind's very existence by overloading the capacity of earth systems to continue their essential functioning.[11] These concerns find expression in the present-day currency of such terms as sustainability and ecological footprints.

---

[10] Paul R. Ehrlich and Anne H. Ehrlich, *Healing the Planet: Strategies for Resolving the Environmental Crisis* (Reading, MA: Addison-Wesley, 1991), 6, 238. See also, e.g., Waechter, *Dessine-moi une planète*, 39: "we must admit that we are living beyond our means."

[11] Donella H. Meadows, Dennis L. Meadows, Jørgen Randers, and William W. Behrens III, *The Limits to Growth: A Report to the Club of Rome's Project on the Predicament of Mankind* (New York, NY: Universe Books, 1972); The Editors of *The Ecologist* (Edward Goldsmith and others), *A Blueprint for Survival* (Harmondsworth: Penguin, 1972).

Some environmentalists offer a third argument for restraining the human alteration of the earth's surface. As well as being the only habitat in which we can biologically exist, they argue, it is the only one in which we can psychologically flourish, as the one in which we ourselves developed as a species. Human beings have deep-rooted needs, whether affective, aesthetic, cognitive, or spiritual, that an environment too much tamed and transformed from its natural state can no longer satisfy. The ecologist Hugh Iltis in 1974 reacted vehemently to the suggestion by an urban planner that plastic trees might more effectively provide many of the benefits of live ones. "There can never be a healthy humanity, both physically and socially," Iltis wrote, "without its ancient evolutionary and ecological basis." Edward O. Wilson's suggestion that "biophilia," or an affinity for other species, is an innate human characteristic is another expression of the idea. As the human race can for a time exceed carrying capacity, such writers would grant, so too it can destroy the foundations of its own deepest happiness, but in both cases not without suffering for it in the end. If the argument is different, the prescription that follows from it is the same: restraining and minimizing human impact for the benefit of people themselves.[12]

If such is environmentalism, anti-environmentalism can take a number of forms. One is the denial, tacit or explicit, that their biophysical surroundings are still significant factors in people's lives, a stance that the sociologists Catton and Dunlap in 1978 called "human exceptionalism" and traced to the assumption that the growing power of technological and social forces has made natural ones comparatively unimportant.[13]

---

[12] Martin H. Krieger, "What's Wrong with Plastic Trees?," *Science* n.s. 179, #4072 (1973), 446–455; Hugh H. Iltis, "Can One Love a Plastic Tree?," *Bulletin of the Ecological Society of America*, 54, #4 (1973), 5–7, 19 (quotation from 7). The idea is extensively developed in the writings of Paul Shepard: see, e.g., *Nature and Madness* (San Francisco, CA: Sierra Club Books, 1982); *Coming Home to the Pleistocene* (Washington, DC: Island Press, 1998); and *Encounters with Nature* (Washington, DC: Island Press, 1999). Roderick Frazier Nash, *Wilderness and the American Mind*, fifth edition (New Haven, CT: Yale University Press, 2014), Chap. 13, discusses some of its other exponents. Waechter, *Dessine-moi une planète*, argued that an artificial world could never match for human beings the stimulating diversity and complexity of the natural one: "A denatured world would be full of tedium." (156). On biophilia, see Edward O. Wilson, *Biophilia* (Cambridge, MA: Harvard University Press, 1984) and Stephen R. Kellert and Edward O. Wilson, eds., *The Biophilia Hypothesis* (Washington, DC: Island Press, 1993).

[13] William R. Catton, Jr. and Riley E. Dunlap, "Environmental Sociology: A New Paradigm," *American Sociologist* 13, #1 (1978), 41–49.

Another is the belief that the terrestrial environment is too vast, mighty, and robust to be seriously destabilized by any tampering mere people are capable of. A third is the simple denial that human action, in fact, has had or is having harmful consequences, whether in the aggregate or in particular cases. But anti-environmentalism forms the clearest antithesis to environmentalism when it turns the latter's presumption in favor of the natural state of things on its head. It presumes, on the contrary, that human-induced environmental change normally represents—at least when undertaken in the right way—an improvement.

For nature does not know best, the argument goes, people do; indeed, only people can be said to know anything at all. Nature is blind and irrational; human beings alone possess intelligence and purpose. A vast scope, then, exists for applied human intelligence to reform the patterns of the earth's surface, which arose through the chaotic operation of mindless non-human forces. People, indeed, being rational, have not only the capacity but also the duty to reconstruct the earth along much more satisfactory lines. Accordingly, whereas environmentalists view the applied sciences of nature, engineering in particular, with suspicion, environmentalists' adversaries see them as the principal agencies of improvement. We can cope with unintended consequences as they occur, they expect, and ever better as our technologies for doing so improve; we can learn with practice, moreover, to anticipate and forestall them; finally, the argument that such consequences occur is not necessarily an argument against change, for they may at least as often be beneficial as harmful. Applied intelligence can enlarge the natural limits that environmentalists emphasize and vastly increase the carrying capacity of an earth that has been reorganized and made more efficient and productive in meeting human wants. Some argue that the ultimate finiteness of the resources that the earth can provide is the very reason why it needs to be remade in order to develop and maximize for human use what is to be had, a line of thought one might call conservationist (for historical reasons to be elaborated in Chap. 2). Or human ingenuity is so powerful that its application can expand carrying capacity enormously and make the planet's resources, and those available in the universe beyond, literally infinite, a line of reasoning better termed "cornucopian."[14] In either case, where environmentalists urge living within nature's limits, their opponents instead recommend enlarging them. And human ingenuity, the latter propose, can furnish superior substitutes for any psychological satisfactions we get from nature—simulated forms of "virtual nature" or other artificial means, such as Krieger's plastic trees—that are more effective (and cost-effective) than the original. Here too, creative human alteration is likely to be better than preservation because it is intelligent where nature is not.

Environmentalists and anti-environmentalists, in short, differ most centrally in their attitudes toward what a famous symposium held in 1955 called "man's role in changing the face of the earth."[15] Environmentalists regard that role chiefly with trepidation and skepticism, and their opponents with high hopes. The latter, in the words of the political scientist John Dryzek, equate "human progress" with "the ability of humans to manipulate the world in ever more effective fashion." "In their more extreme moments," Dryzek continued, they "believe that a total control of nature is within our grasp."[16] They regard the advent of what is now called the Anthropocene, an epoch of earth history where human forces have attained sufficient size and power to play a major role in reshaping the global environment, as an important step forward and the realization of a longtime hope. They would accept another widely used term, a "good Anthropocene," as an apt label for a future in which those forces are deployed with maximum effect to improve upon nature.

It is time to give them a name of their own, and I will follow Dryzek (and some other writers) in calling them environmental Prometheans, for reasons he succinctly stated: "In Greek mythology Prometheus stole fire

[14] I here diverge somewhat from John S. Dryzek, *The Politics of the Earth: Environmental Discourses*, 3rd ed. (Oxford: Oxford University Press, 2013), who described a cornucopian belief in the unlimited abundance of nature (when properly manipulated by human intelligence) as a component of what he and I call Prometheanism. Yet, he himself quoted even the Promethean Julian Simon's disavowal of the label "cornucopian" (52, 59–60). Prometheanism is best understood as a belief in the desirability of the radical reform of nature by human beings, resulting in either an infinitude or merely an increased abundance of resources. "Cornucopians" are those who expect the former, but those envisioning the latter are equally Promethean in seeing nature's essential finitude as an urgent reason for seeking all that rational human management can make it provide.

[15] William L. Thomas, Jr., ed., *Man's Role in Changing the Face of the Earth* (Chicago, IL: University of Chicago Press, 1956).

[16] Dryzek, *The Politics of the Earth*, 61, 64.

from Zeus, and so vastly increased the human capacity to manipulate the world." Fire is both an important tool of environmental transformation in itself, what Omer C. Stewart in the "Man's Role" symposium called "the first great force employed by man" in that task, and a symbol for all other such tools that human beings have applied to the work of engineering the planet.[17] Prometheanism is a form of environmental, though not of environmentalist, thought. It is environmental in the profound importance it attributes to the biophysical surroundings of human life. (If they do not matter, why bother to alter them?) It is anti-environmentalist in that it rejects environmentalism's skepticism about the human capacity to alter them fundamentally and for the better. The Promethean credo corresponds to what Paul Wapner defined as "the dream of mastery" as that of environmentalism does to his "dream of nature." In his words, it "suggests that humans will realize their highest potential as individuals, societies, and species to the degree that we can manipulate the natural world as we see fit." Drawing much the same distinction, Clive Hamilton has described a similar divide in environmental matters between technologically hubristic "Prometheans" and more skeptical "Soterians," whom he names after "Soteria, the goddess of safety, preservation and deliverance from harm."[18]

"Promethean" is not the only available name. Others have called the same outlook a "Faustian" one, invoking a legendary scholar who sought knowledge for the sake of power over nature. In the second part of his dramatic poem *Faust*, published in 1832, J. W. Goethe showed the title character pursuing such mastery, not merely in the study or the laboratory, but on the earth's surface in a great feat of environmental engineering. With the devil's assistance, Faust attempts to reclaim by drainage a

[17] *Ibid.*, 52; Omer C. Stewart, "Fire as the First Great Force Employed by Man," in Thomas, *Man's Role*, 115–133.

[18] Wapner, *Living Through the End of Nature* (quotation from 104; see also his Chap. 4, "The Dream of Mastery," 79–105); Clive Hamilton, *Earthmasters: The Dawn of the Age of Climate Engineering* (New Haven, CT: Yale University Press, 2013), 18. It is true that Wapner described both perspectives as increasingly outmoded, but his discussion testifies to their past importance and to their stubborn persistence today. For Hamilton, the distinction is, if anything, more meaningful than ever. The categories both use parallel in some ways the influential distinction drawn by Donald Worster between "arcadian" and "imperial" ecologists, beginning with the eighteenth-century precursors of the science: *Nature's Economy: A History of Ecological Ideas* (New York, NY: Cambridge University Press, 1985).

large tract of "wasteland," a coastal marsh, for people to settle and cultivate.[19] But the very example suggests why the label is better avoided. Rarely, if ever, do we call something that we approve of "Faustian." The brutal insensitivity of Faust's reclamation scheme (the violence it inflicts on the people living in its path) is too glaring, and the source of his powers too explicitly diabolical, for his name to make a neutral term of description. It does admirably for polemics but not for impartial analysis.

The older myth of Hercules offers another possibility. To a number of nineteenth-century writers, Hercules represented the virtues of work and labor in general and of the conquest of nature in particular. The Greek hero, they noted, had smothered or enslaved the natural monsters of the ancient world, conquered rivers and compelled them to do his work, forced aside two worlds in order to connect the Mediterranean to the great ocean beyond, pierced mountains, purged marshes, cleared forests, and "pacified and civilized nature." As a result, he had aroused the antagonism of the old deities of the earth, who clamored to Jupiter for his death.[20] But the same failing that disqualifies Faust as a symbol disqualifies Hercules, too, though in the opposite way. His name does not make a neutral label. Unlike Faust's, it carries too heavily positive a set of connotations. Rarely do we call something we fundamentally disapprove of "Herculean." A much more mixed set of suggestions clusters around "Promethean" today. Some writers use it to name a position they attack, others one that they defend.[21] One can use it without thereby prejudging matters.

---

[19] For environmental readings of the conclusion of Goethe's *Faust*, see Bruce Rich, *Mortgaging the Earth: The World Bank, Environmental Impoverishment, and the Crisis of Development* (New York, NY: Earthscan, 1994), 104–106; Hans Christoph Binswanger, "The Challenge of Faust," *Science* n.s. 281, #5377 (1998), 640–641; and Hans C. Binswanger and Kirk R. Smith, "Paracelsus and Goethe: Founding Fathers of Environmental Health," *Bulletin of the World Health Organization* 78, #9 (2000), 1162–1164.

[20] Jules Michelet, *Bible de l'humanité* (Paris: F. Chamerot, 1864), 227–230, 238–239 (quotation from 239); see also Pierre-Simon Ballanche, *Essais de palingénésie sociale: Tome premier: Prolégomènes* (Paris: Jules Didot aîné, 1827), 71–72 and André Poëy, *Le positivisme* (Paris: Librairie Germer-Baillière, 1876), 160.

[21] Compare, for example, Dryzek's largely unfavorable use of the label with its adoption by Martin W. Lewis in his *Green Delusions: An Environmentalist Critique of Radical Environmentalism* (Durham, NC: Duke University Press, 1992).

"Environmentalist" and "Promethean" as I have defined them are not the either-ors of a dichotomy, but the poles of a continuum. One is likelier to encounter a pure, uncompromising example of either in caricatures by the opposing side than in flesh and blood. Few people would wish all of the changes that people have brought about in the earth's surface unmade. Few would wish it changed further into an entirely and unmistakably human-made landscape. But the distinction between the two schools of thought corresponds to a real divide as to what should be done now. Environmentalists hold that human-induced environmental changes have already gone too far for our own safety. Prometheans disagree. Just as the "end of nature," even if granted, is by no means fatal to the distinction, neither is the fact that the question sometimes now arises (regarding climate change, e.g.) of intervening in nature to "preserve" it. The paradox for environmentalists is only an apparent one. On their principles, the wisdom of acting or not acting in certain ways would largely depend on which choice offered the least risk of further disruption and the least net interference (including, of course, they would emphasize, the dangers of unforeseen side effects that even well-intentioned action might produce). Environmentalists might well reluctantly conclude today that some forms of geoengineering offer the best chance of stabilizing a global climate threatened by other human actions that show no signs of being brought under control. Unlike Prometheans, they would not look forward to the possible "improvements" in the climate that geoengineering techniques might eventually make possible.

One reason that the imagery of "spaceship earth," despite a promising start in the 1960s, has not become more common than it has may lie in the awkward way that it cuts across the environmentalist/Promethean divide. For environmentalists, it has the appeal of dramatizing limits and the need for expert direction to avoid dangers that threaten all of those aboard. Yet, they would warn that thinking of the earth as a spacecraft—something that we have made and whose workings some of us, at least, thoroughly understand—suggests far more knowledge of the biosphere than they would grant we possess or ever will possess, and fosters far more confidence in our ability to tinker with it than we should feel. Nothing could be less environmentalist or more Promethean in its deeper implications than a title like R. Buckminster Fuller's *Operating Manual for*

Spaceship Earth—or, for that matter, more Promethean than the book's confident invocation of the power of human ingenuity.[22]

Prometheanism has a history and a philosophy, but neither historians nor philosophers have paid them nearly as much attention as they have paid to environmentalism's. Precursors or exponents of green thought have been studied much more carefully (and sympathetically) than have prophets of human mastery over the earth. This is not to say that Prometheanism's existence has been denied or ignored; quite the contrary. It has often been supposed, rather, to have been a part of the common climate of opinion, so pervasive and unquestioned before about the mid-twentieth century that individual statements of it seemed not worth examining in detail, or not even worth looking for. Dryzek, who has provided the fullest summary of its present-day components, described it as having been, for some three centuries before the rise of modern environmentalism (i.e., until around 1960), a dominant, but at the same time, an unarticulated earthview, one of those discourses "so ingrained and taken-for-granted that it would never occur to anyone to mention them," let alone defend them intelligently and in detail.[23] Yet, as a matter of historical fact, some people did state and defend an environmental Prometheanism long before 1960. Who were they? What did they say in its support? With what beliefs about politics, economics, and society did they couple their environmental doctrines?

It is true that the Prometheans of the past lacked a corps of vocal opponents who could provoke them into arguing back. Therein lies the large kernel of truth in what Dryzek has said. Most people in the past few centuries in the Western world whose thoughts have left a trace do, indeed, seem to have held more or less Promethean presumptions, at least tacitly, and very few those of present-day environmentalism. But even if we were

---

[22] R. Buckminster Fuller, *Operating Manual for Spaceship Earth* (New York, NY: E. P. Dutton, 1978). Joachim Radkau, *The Age of Ecology: A Global History*, trans. Patrick Camiller (Malden, MA: Polity Press, 2014), 111, aptly describes the imagery of "spaceship earth" as characteristic of the 1960s rather than the present.

[23] Dryzek, *Politics of the Earth*, 52–53. "It is undeniable, historically, that Marxism includes a triumphalist view of 'man's conquest of nature,'" Raymond Williams also wrote, but this, he argued, represented less a feature characteristic of Marxism than its infection by attitudes common to nineteenth-century thought in general: "Problems of Materialism," *New Left Review* #109 (May-June 1978), 8–9.

all Prometheans then, some were more Promethean than others. There is a difference, one of degree if not of kind, between passively accepting a credo and actively expounding it. Full-throated Prometheans, it seems reasonable to assume, took their environmental beliefs more closely to heart than more reticent ones did. If we want to know more about Prometheanism's past political associations, we can at least look to the political stances of those who preached it most fervently and intelligently.

By what markers can we identify them? There are some pitfalls to avoid. The largest and most dangerous surround the word "nature" and the various meanings that it possesses. The English statesman-philosopher Francis Bacon (1561–1626), Prometheanism's greatest early prophet, sidestepped one pitfall in his famous axiom that "we cannot command nature except by obeying her."[24] He realized, that is to say, that a certain attitude towards some parts of what we refer to as "nature" does not exclude a quite different attitude towards others. Environmental Prometheanism is a case in point. Nature understood as a set of processes, particularly ones described in laws—if this happens, then this does—is not the same as nature as a set of states or configurations of matter. Environmental Prometheans do not seek to alter nature's most basic laws. (Other kinds of Prometheans may, but this book is not about them.) Rather, they propose, like Bacon, to use those laws to alter nature's patterns and products. What they seek to alter would (were usage not already firmly fixed) be much less confusingly termed physical geography—the landforms, weather and climate, soils, flows and bodies of water, flows of energy and matter, plant and animal life, land cover, and chemical composition of the earth's surface—than nature.

Similarly, if one takes "nature" as an umbrella term for the whole universe of reality, which includes human beings and their doings (in John Stuart Mill's words, "all the powers existing in either the outer or the inner world and everything which takes place by means of those powers"), attitudes toward it in no way distinguish environmentalists from Prometheans. It would be meaningless then to talk of changing nature, or improving it, or deferring to it, or damaging it, for all actions one

[24] James Spedding, Robert Leslie Ellis, and Douglas Denon Heath, eds., *The Works of Francis Bacon*, vol. 4 (London: Longman & Co., 1858), 114.

takes are within nature and a part of it.[25] But again, if one recalls that environmental Prometheans wish to change physical geography rather than "nature" in this sense, the distinction retains its usefulness. Choices genuinely exist between altering and not altering, or much altering versus little altering, a part or the whole of the earth's surface.

The frequent use of "nature" to denote things other or broader than its chief meaning for environmentalists—physical geography little modified by human action—is a fertile source of confusion. The historians Lovejoy and Boas observed in 1935 "that the word 'nature' is probably the most equivocal in the vocabulary of the European peoples; that the range of connotations of the single term covers conceptions not only distinct but often absolutely antithetic to one another in their implications." The "extreme multiplicity of meanings" that they found even in its cognates in classical antiquity has certainly not diminished in our day.[26] Writers who praise "nature," perhaps meaning the term in any of a number of other senses than that of physical geography, are not necessarily green or environmentalist, for they may at the same time, and without contradicting themselves, express wholehearted approval for Promethean projects of reshaping the earth's surface. An enthusiasm for the development of renewable energy or renewable resources in not necessarily or inherently an environmentalist one, for it may be tied to a strongly Promethean program of land-cover transformation to harness them for use. Nor, for example, though we may tend to think of grass and foliage as pieces of "nature," does the prevalence of lawns and gardens and trees in the ideal landscape described by a utopian prophet necessarily signal an environmentalist as opposed to a Promethean vision. They may be green herrings, as it were. Indeed, environmentalists consider the American suburban lawn, a monoculture of a single introduced species maintained by heavy inputs of labor, energy, and chemicals, to be anything but a model land cover.[27] Such manipulating, recombining, and

---

[25] John Stuart Mill, "Nature," in *Three Essays on Religion: Nature, The Utility of Religion, and Theism* (London: Longmans, Green, Reader, and Dyer, 1874), 8.

[26] Arthur O. Lovejoy and George Boas, *Primitivism and Related Ideas in Antiquity* (Baltimore, MD: Johns Hopkins University Press, 1935), 11–12.

[27] See, e.g., Ted Steinberg, *American Green: The Obsessive Quest for the Perfect Lawn* (New York, NY: W. W. Norton, 2006).

relocating elements of the environment are a far cry from the deference that environmentalists propose. So is picking and choosing which among them is useful and worth keeping, rather than assuming a usefulness that we merely do not understand.

Another possible criterion should be used with great caution. Environmental Prometheans need not regard all human-induced changes as beneficial and may express misgivings or even alarm over many that have occurred. (Indeed, one could hardly imagine a person who welcomed any and all conceivable changes, even ones the reverse of one another.) What characterizes Prometheanism, rather, is the belief that human redesign can improve immensely and profoundly on nature *if* society itself is properly organized to carry out the work, whereas environmentalists see most such hopes as illusory and the project as inherently dubious. No one holding strong political views would expect the wrong kind of regime to produce the right kind of Promethean change, but that kind of skepticism should not be mistaken for skepticism about the promise of Promethean change itself. In *The Communist Manifesto*, Karl Marx praised among the achievements of the capitalist stage of economic development its unprecedented subjugation of the earth's surface to human wants, its "clearing of whole continents for cultivation, canalization of rivers, whole populations conjured out of the ground."[28] He was no less a Promethean, because, in several other texts, much cited today by his admirers, he criticized capitalism for in many ways doing the job badly and inflicting unnecessary damage. In Dryzek's words, Marx all the same looked forward to a "future beyond capitalism" featuring "technological progress, economic growth, and the conquest of nature."[29] It follows that many concerns of the past and present that, by other definitions, are often called environmentalist—many regarding pollution, for example—are, by mine, environmental but not environmentalist if they do not fundamentally question the Promethean project as such.

---

[28] Karl Marx, *The Communist Manifesto*, ed. Frederic L. Bender (New York, NY: W. W. Norton, 1988), 59.

[29] Dryzek, *Politics of the Earth*, 53; see also Williams, "Problems of Materialism"; Val Routley, "On Karl Marx as an Environmental Hero," *Environmental Ethics* 3, #3 (1981), 237–244; Ted Benton, "Marxism and Natural Limits: An Ecological Critique and Reconstruction," *New Left Review* #178 (1989), 51–86.

Prometheanism is certainly not a tacit or unarticulated discourse today, for it now has many vocal environmentalist critics to defend itself against. One tends, as already noted, to find its contemporary defenders on the political right, just as one tends to find its critics on the left. Dryzek particularly associated it with free-market conservatism, one of whose American idols, Ayn Rand, wrote in 1971: "In order to survive, man has to discover and produce everything he needs, which means that he has to *alter* his background and adapt it to his needs. … It is not merely symbolic that fire was the property of the gods which Prometheus brought to man. The ecologists are the new vultures swarming to extinguish that fire."[30] So, too, Llewellyn Rockwell, founding chairman of the Ludwig von Mises Institute and long the foremost promoter in the USA of its free-market Austrian economics. "How glorious," he wrote, echoing a passage from St. Augustine he had just quoted, "to see human habitations spreading where once unchecked nature reigned. … All swamps should be drained, all rain forests turned over to productive agriculture. … From time immemorial until the day before yesterday, Western man has seen nature as the enemy, and rightly so. It is dangerous and deadly. For the sake of our own survival, it must be tamed, cut, curbed, controlled," whereas an environmentalist would say that, for the sake of our own survival and physical and mental health, it is rather our impacts on nature that must be tamed, cut, curbed, controlled.[31] The neoliberal economist Julian Simon argued forcefully that the resources provided by the natural world are not limited and become more, not less, abundant as humankind takes them in hand for systematic management, so long as it allows private ownership, market forces, and technological innovation to operate with minimal interference. Simon made a famous bet with the biologist and environmentalist Paul Ehrlich and colleagues in the late twentieth century on trends in nonrenewable resource prices. He won

---

[30] Ayn Rand, *Return of the Primitive: The Anti-Industrial Revolution*, ed. Peter Schwartz (New York, NY: Meridian, 1999), 277.

[31] Llewellyn Rockwell, *The Left, the Right & the State* (Auburn, AL: Ludwig von Mises Institute, 2008), 153, 154.

when all of the five metals chosen as test cases became cheaper over the course of a decade.[32]

And environmental Prometheanism has found advocates on the traditionalist side of the American right as outspoken as on the libertarian or free-market one. It seems not to be one of the issues, such as immigration and abortion, on which the alliance of the two wings occasionally threatens to come apart. As Dryzek observed, it "resonates with the interests of both capitalist market zealots and Christian conservatives."[33] One of the forms it takes for the latter is "dominion theology," invoking the Bible's grant to human beings of authority over the earth and its animals and plants. Though support for environmental causes has grown in recent years among American fundamentalist or evangelical Christians, there remains proportionally less of it than there is in the population overall or among mainline Protestants and Roman Catholics.[34] The Cornwall Alliance for the Stewardship of Creation issued a declaration in 2000 stating a moderate but firm Christian Prometheanism. Radical environmentalists, it asserted, "ignore our potential, as bearers of God's image, to add to the earth's abundance." It rejected the thesis that "nature knows best" as tending "to deify nature or oppose human dominion over earth." It concluded by stressing people's duty "to join with God in making provision for our temporal well being and to enhance the beauty and fruitfulness of the rest of the earth." Such notable American social conservatives and neoconservatives as James Dobson, Elizabeth Fox-Genovese, Peter Huber, Deepak Lal, Herbert London, and Richard John

---

[32] Paul Sabin, *The Bet: Paul Ehrlich, Julian Simon, and Our Gamble over Earth's Future* (New Haven, CT: Yale University Press, 2013) has offered the most detailed history of the bet as an expression of Simon's resource Prometheanism. See also Mayhew, *Malthus*, 205–212.

[33] Dryzek, *Politics of the Earth*, 67. Peter J. Jacques, in *Environmental Skepticism: Ecology, Power and Public Life* (Farnham, Surrey: Ashgate Publishing Limited, 2009), has likewise (1, 10, 71) described right-wing evangelical Christianity as a central constituency of modern anti-environmentalism.

[34] Stephen R. Rock, *Faith and Foreign Policy: The Views and Influence of U.S. Christians and Christian Organizations* (New York, NY: Continuum, 2011), 162–166. As he observed, Prometheanism is not the only environmental belief system on the American Christian right. Two others are "end times" theology, which sees the disasters being brought about by human action as signs that the biblically prophesied end is near, and a belief that a divinely designed earth is too robust to be seriously harmed by human action: *ibid.*, 169, 171.

Neuhaus signed the declaration.[35] Its principal draftsman, the evangelical theologian E. Calvin Beisner, further developed its Promethean message in other works. He criticized attempts to soften the Biblical message of human dominion over the earth as contrary to the Hebrew words in question, which he thought best translated as "subdue" and "rule" and as "conveying strong, forceful subjugation."[36] He included among humankind's tasks in redeeming its own fallen nature and an earth cursed by its fall those of "multiplication, subduing and ruling the earth, transforming the wilderness by cultivation into a garden, and guarding that garden from harm," not least from harm threatened by the agencies of nature itself.[37] Natural resources, Beisner argued, drawing heavily on the work of Julian Simon, have no limits; human beings make them steadily more abundant by using the creativity that reflects, however imperfectly, that of their creator.[38] He combined these arguments with warnings about the "statist oppression" that environmental regulations might entail, and his other writings have included attacks on such targets as economically redistributive policies and "the multiculturalist bandwagon."[39]

While not outspokenly religious, Europe's most prominent anti-green politician of recent years, Vaclav Klaus, Prime Minister and later President of the Czech Republic, has combined an environmental Prometheanism with a right-wing ideology uniting free-market doctrines in economics with a traditionalistic and nationalistic stance in other matters. So, too, the "Wise Use" movement that developed in the western USA in the late twentieth century has called at once for greater privatization and development and reduced regulation of the public domain and for the continuance of traditional modes of land use. Its chief ideologist assailed

---

[35] Cornwall Declaration at http://www.cornwallalliance.org/2000/05/01/the-cornwall-declaration-on-environmental-stewardship/; signers' names from http://www.cornwallalliance.org/1999/10/29/notable-signers-of-the-cornwall-declaration/.

[36] E. Calvin Beisner, *Where Garden Meets Wilderness: Evangelical Entry into the Environmental Debate* (Grand Rapids, MI: Acton Institute for the Study of Religion and Liberty and W. E. Eerdmans, 1997), 12–14, 103, 178n21.

[37] Beisner, *Where Garden Meets Wilderness*, 25.

[38] Beisner, *Where Garden Meets Wilderness*, 25, 60–61, 63–65, 102, 108–109, 167.

[39] Beisner, *Where Garden Meets Wilderness*, 158; *Prosperity and Poverty: The Compassionate Use of Resources in a World of Scarcity* (Westchester, IL: Crossway Books, 1988); "The Double-Edged Sword of Multiculturalism," *The Freeman* 44, #3 (1994), 104–112 (quotation from 110).

environmentalism as representing simultaneously a "radical" philosophy undermining the heritage of the human past and an obstacle to the Promethean mission of "improving the earth for the massive use of future generations." Efforts towards such improvement, he granted, at times create problems that have to be met through adaptive learning. "People's reworking of the earth," though, he concluded, is "ultimately benevolent." "To Wise Users," as a sympathetic analyst of the movement has written, "nature … was imperfect until 'improved' by human labor," a classically Promethean belief.[40]

Dryzek has accurately catalogued the political and economic beliefs most often associated with environmental Prometheanism today, notably an ideology of free markets and minimal government, a neoliberal capitalism grounded chiefly in orthodox economics and reliant on market competition as the driving force of progress: a credo that has not proved at all repugnant to equally anti-government conservatives of a principally traditionalistic bent. "Inasmuch as they attend to government, Prometheans see mainly sources of ill." Another writer has similarly described Prometheanism as "a powerful strand of Western technological thinking and conservative politicking that sees no ethical or other obstacle to total domination of the planet." Nonetheless, these observed affinities, a right-wing politics mingling a liking for the market with a hostility to the state, should not be made a part of the *definition* of Promethean thought. Prometheanism is recognizable by its project of the reshaping of the earth for the better by human minds and hands, by, in Dryzek's words, its "forthright and confident celebration of human domination of nature," and not by any particular means, such as markets or private property, that its advocates in certain times and places may have supposed will promote or frustrate that end.[41] We would better understand its present-day political associations, how far they are inherent and

[40] Ron Arnold, "Overcoming Ideology," in Philip D. Brick and R. McGregor Cawley, eds., *A Wolf in the Garden: The Land Rights Movement and the New Environmental Debate* (Lanham, MD: Rowman and Littlefield, 1996), 23, 24. Chase, *In a Dark Wood*, 380; see also James Morton Turner, "'The Specter of Environmentalism': Wilderness, Environmental Politics, and the Evolution of the New Right," *Journal of American History* 96, #1 (2009), 123–148.

[41] Dryzek, *Politics of the Earth*, 61 (quotation), 64, 89 (quotation); Hamilton, *Earthmasters*, 18.

necessary and how far they may be accidental, if we knew more about the company that it had kept in the past.

We can best begin, though, by exploring the logic of today's pattern. If a conservative or right-wing outlook thought tends to accompany Promethean views in environmental matters, what are its own defining features, and how do they explain the association? It would be irresponsible to pretend that there are no difficulties of classifying thinkers, both past and present, according to a schema so one-dimensional and, perhaps, so time- or culture-bound as the ideological continuum running between right and left. Certainly one cannot easily apply it in a meaningful way to thinkers before about 1790, when "right" and "left" as political categories first came into use and when Edmund Burke (1729–1797) wrote what remains by wide agreement the archetypal English-language conservative text, his *Reflections on the Revolution in France*. But that the categories have remained and indeed flourished in use since then, and that people understand them and routinely employ them to characterize their own and others' political stances, suggests that they have a coherence and aptness that the passage of time has not destroyed. What precisely, though, do they mean?

We seem, in practice, to use a number of elements when deciding where to locate someone or some stance on the left-right continuum. In stating the first, and perhaps the single most salient of them, it would be very broadly true to say that the right is averse to change, and the left is not, even without specifying what kinds of change are involved. When the distinction is rephrased as one between innovation and tradition, rather than merely between stability and change, most of the problems with the initial formulation disappear. Conservatism in the most basic sense, as the dictionary will attest, entails a distrust of novelty and a preference for what is tried and established. It favors order, stability, and continuity with the past. It looks skeptically upon projects for reshaping the world, even when a proposed reform seems far more rational than the arrangements it would supplant. Emphasizing limits, dubious about the possibilities of radical improvement in the human condition, conservatism is, on the other hand, alert to the possibilities of radical deterioration, especially through an ill-advised abandonment of ways of life that have stood a long test of time.

Progressivism, defined in very abstract and general terms, is conservatism's ideological antithesis. It attaches much less value to tradition and stability, regarding both less as important virtues than as smokescreens used to hide or justify the inequities and the inefficiencies of the status quo by those who have a stake in its perpetuation. Progressives view the possibilities of improvement in the human condition much more confidently and expansively than conservatives do, and they regard the risks that reforms may entail with much less foreboding. They are apt to think reason, which conservatives often disparage, a more trustworthy guide than tradition, with its large element of unexamined prejudice and error.

Progressives, then, see, and conservatives do not see, much potential for existing ways or ones sanctioned by custom and precedent to be altered for the better. The latter urge a presumption in favor of anything that has the backing of long usage even if fully satisfactory reasons for it cannot be articulated. They are likely to see most of the progress that progressives seek as illusory and unattainable, or as unattainable without costs that are too high. Progressives, for their part, emphasize the dangers and costs incurred and the suffering unnecessarily perpetuated, not by change, but by failure to change and by an excessive veneration of custom and tradition. Enthusiasm for brave new worlds and great leaps forward is more characteristic of the left, and skepticism about them of the right. The admirably even-handed definitions offered by Ambrose Bierce in the late nineteenth-century USA center on this set of oppositions: "Conservative, n.: A statesman who is enamored of existing evils, as opposed to the Liberal who wishes to replace them with others."[42]

This basis for the left/right distinction works well when applied to the French Revolution, the fountainhead of the distinction itself. The two factions in the Revolution (identified according to the seating patterns in the newly established national assemblies beginning in 1789, those who gathered to the right and left of the Speaker) indeed represented different attitudes towards tradition and progress. Their initial and defining divisions occurred over the future of the main pillars of old-regime France: the monarchy and the Church. The left, on both counts, favored a radical

[42] Ambrose Bierce, *The Devil's Dictionary, Tales, and Memoirs*, ed. S. T. Joshi (New York, NY: Library of America, 2011), 463.

break with the past that the right resisted. More generally, for the left: "Revolution meant rejecting the past, introducing a sense of discontinuity in secular time ... shaping the future in accordance with the discoveries made in the present."[43] A cautious and skeptical attitude towards innovation, conversely, pervaded Burke's *Reflections*, the most notable contemporary attack on the Revolution and its doings, which has transcended the occasion that gave rise to it and has become a classic statement of the principles of a moderate and reflective conservatism. Burke was very far from being a reactionary or a violent or mindless opponent of change; he spent most of his parliamentary career in opposition to the conservative dominance of British politics. But his *Reflections* and his subsequent writings on France emphasized the dangers rather than the benefits of political novelty and the preeminent value of established and traditional institutions. They led, indeed, to a break with his former Whig associates.

Leading figures in later English political thought elaborated the same distinction. John Stuart Mill spoke of "the classification which may be made of all writers into Progressive and Conservative." The former, he observed, tended to emphasize "those truths with which existing doctrines and institutions were at variance," the latter "the neglected truths which lay *in* them," or the insights and values that, however inarticulately, they embodied. Mill used Samuel Taylor Coleridge as an example of an intelligent conservative, one for whom anything long-established had a presumptive claim to being useful, while he pointed to Jeremy Bentham as an archetypal progressive, "the great questioner of things established" and the great advocate of rational improvement.[44] The twentieth-century English philosopher Michael Oakeshott approvingly identified conservatism with a certain disposition or outlook: "to prefer the familiar to the unknown, to prefer the tried to the untried." Conservatism, Oakeshott

[43] Nicholas Atkin and Frank Jallett, "Introduction: Les Droites Commençent Ici," in Nicholas Atkin and Frank Jallett, eds., *The Right in France, 1789–1997* (London: I. B. Tauris, 1998), 1–17; John M. Roberts, "The French Origins of the 'Right,'" *Transactions of the Royal Historical Society*, 5th series, 23 (1973), 27–53; Lynn Hunt, "The World We Have Gained: The Future of the French Revolution," *American Historical Review* 108, #1 (2003), 6 (quotation).

[44] Alan Ryan, ed., *Utilitarianism and Other Essays: J. S. Mill and Jeremy Bentham* (New York, NY: Penguin, 1987), 133, 177.

continued, "is averse from change, which appears always, in the first place, deprivation," and it finds "small and slow changes more tolerable than huge and sudden"; the conservative, it follows, will not be "an ardent innovator."[45] Among Americans, Ralph Waldo Emerson described "[t]he two parties which divide the state" as those of "Conservatism" and "Innovation" (his strongest sympathies, unlike Oakeshott's, lying plainly with the latter).[46] The long persistence of the Revolution itself and of its main targets as points of political reference in France testifies to the salience of the same division. Pre-1917 Russian conservatism likewise centered on the maintenance of the old political and social order, based on tsarist autocracy and the Orthodox Church.[47]

It is this criterion, though, that makes the entente between traditional conservatives and free-market classical liberals, the two wings of the contemporary Western political right, something of a puzzle. The Austrian-born economist Friedrich Hayek, the latter group's ablest twentieth-century spokesman, explicitly declared himself a liberal (in the classical European sense) and not a conservative. He did not, he made it clear, share the conservative's automatic distrust of change, reverence for tradition, or skepticism about the possibility of progress, though he granted that the pressures of history had made conservatives for the time his allies against a common enemy that he identified as socialism.[48] But Hayek's own writings on politics do much to explain the ability of cultural and free-market conservatism to coexist, however uneasily. They express a deep aversion to rational state planning, one shared, albeit for somewhat different reasons, by traditionalists. Hayek emphasized the value of social orders, notably ones involving private property and free markets, that develop not through deliberate design, but rather by a kind of Darwinian social selection as they enrich and empower those societies that retain

[45] Michael Oakeshott, "On Being Conservative," in *Rationalism in Politics and Other Essays* (London: Methuen & Co., 1962), 169, 170, 171.

[46] Ralph Waldo Emerson, "The Conservative," in *The Collected Works of Ralph Waldo Emerson, vol. 2: Nature, Addresses, and Lectures*, ed. Robert E. Spiller (Cambridge, MA: Belknap Press of Harvard University Press, 1971). 184.

[47] Richard Pipes, *Russian Conservatism and Its Critics: A Study in Political Culture* (New Haven, CT: Yale University Press, 2005).

[48] Friedrich Hayek, "Why I Am Not a Conservative," in *The Constitution of Liberty: The Definitive Edition*, ed. Ronald Harmowy (Chicago, IL: University of Chicago Press, 2011), 519–533.

them relative to those that reject them. In Hayek's eyes, their replacement by supposedly more rational planned orders, usually by a state overseeing the ownership and distribution of goods, represents the principal threat in the modern world to the prosperity, freedom, and human well-being that a spontaneously evolved order best provides.[49] What Hayek called spontaneously evolved orders, cultural conservatives today are likely to respect and to uphold as traditions, and indeed Hayek, without himself being a believer, thought one important set of traditions, those of religious practice, good for other people to have, as carriers (at least in Western Christendom) of some of the values essential to the maintenance of a classical liberal society.[50]

That puzzle is more easily resolved than another, which is central to the concerns of this book. If the defining elements of the political right include skepticism about change, especially profound and rapid change, and an attachment to stability and continuity, why does one now find Promethean proponents of environmental transformation, of profound and rapid reform of the physical world, chiefly in its ranks? And the puzzle is not merely one of right-wing Prometheanism, but also of left-wing environmentalism. Why do those who minimize the dangers of innovation and emphasize the opportunities for reforming the social status quo not do so too when it is the natural status quo that is in question? Why do such kindred hallmarks of conservative social rhetoric as order, limits, the virtues of restraint, and the distrust of utopian hopes for "changing the world" reverse themselves as soon as the conversation turns to the prospect of changing the earth? Why does the progressive, Mill's "questioner of things established," refrain from questioning things established in physical geography, and why has "progress" become almost a tabooed term for progressives themselves in environmental matters—but only there? Promethean environmental transformationism, found, as Dryzek observed, chiefly on the contemporary right, seems little compat-

[49] For a succinct presentation, see Friedrich Hayek, *The Fatal Conceit: The Errors of Socialism*, ed. W. W. Bartley III (Chicago, IL: University of Chicago Press, 1989); for a fuller one, F.A. Hayek, *The Market and Other Orders*, ed. Bruce Caldwell, *The Collected Works of F. A. Hayek*, v. 15 (Chicago, IL: University of Chicago Press, 2014).

[50] Hayek, *The Fatal Conceit*, 135–140.

ible with the classic watchwords of conservatism—and indeed with the term conservatism itself—but highly consistent with those of the left.

To observe, as the green political philosopher Andrew Dobson has done, that there is a key difference—present-day conservatives value and wish to preserve the legacies of human history, progressives those of natural history—is largely true as a matter of fact, but it is not an explanation.[51] It leaves unanswered the key question: on what grounds do conservatives and progressives thus differ? Would it not, after all, be more philosophically and rhetorically consistent for each to value or to discount both kinds of heritage together? Prometheanism is radical in the root sense of that word, proposing fundamental and sweeping environmental change, and yet its advocates are the opposite of what popular parlance labels radicals. Likewise, environmental restoration, or the reversal (rather than merely the stopping or slowing) of human-induced change, is the environmental analogue to a reactionary politics, farther right than mere conservatism, but few of those whom we would call political reactionaries champion it. We seem here to have found, not the reason why environmentalists and Prometheans subscribe to the politics that they do, but rather a reason for expecting the opposite.

As noted earlier, a classic criticism of any proposed change is that, by disrupting a complex system that has developed over a long time, it will bring about an array of unintended and undesirable consequences. Few arguments are more characteristic of the right in its critiques of proposed social reforms. Indeed, A.O. Hirschman in 1991 identified three variants of the unintended-consequences critique as constituting the "rhetoric of reaction." In attacking the anti-poverty programs of the modern welfare state, for example, Hirschman observed, social, political, and economic conservatives deployed all three variants. The "futility thesis" holds that attempts to abolish poverty will typically fail to make a significant difference to a problem whose roots are too deep for mere laws or transfers of income to reach. The "perversity thesis" holds that such attempts will, in fact, make things worse: that generous welfare systems, for example, erode the work ethic and create a culture of dependence, worsening and entrenching the very problems of poverty and despair that

[51] Andrew Dobson, *Green Political Thought*, 4th edition (London: Routledge, 2007), 161–162.

they meant to cure. The "jeopardy thesis" holds that, even if successful in their intended goals, they will endanger something else of great value, as the political legitimacy of the state, once it launches generous welfare programs, is undermined by its inability to meet the public's ever-more insatiable expectations for benefits. The three forms of the argument unite in the conclusion that they reach, one basic to conservatism: that deliberately and rationally effecting progress and changing the world for the better is a difficult, dangerous, and in many cases impossible task, whereas inadvertently changing the world for the worse in trying to improve it, if only through the waste of misapplied resources, is not difficult at all.[52] In his statement of the case for conservatism, Michael Oakeshott put much weight on the way in which any change "entails certain loss and possible gain." Not only does it destroy the stability of things as they were, but "[t]he total change is always more extensive than the change designed ... whenever there is innovation there is the certainty that the change will be greater than was intended" because of the ubiquity of unanticipated consequences.[53]

Again, the present-day politics of the environment confound the expectations that one might form on this basis. Few arguments are more characteristic of environmentalism, as already noted, than that of the danger of undesirable consequences produced by meddling in the complex system of nature, exactly the argument conservatives so often make with respect to the complex system of society. The prudence and aversion to risk that are characteristic of conservatism in other realms are, in Andrew Dobson's words, "'the Greens' 'precautionary principle' for decision-making in all but name."[54] Carson's *Silent Spring* developed what Hirschman would later dub the "jeopardy thesis" in the area of insect pest management, arguing that the single-minded pursuit of a narrow and seemingly quite desirable goal had placed many valuable com-

[52] Albert O. Hirschman, *The Rhetoric of Reaction: Perversity, Futility, Jeopardy* (Cambridge, MA: Belknap Press of Harvard University Press, 1991).

[53] Oakeshott, "On Being Conservative," 172.

[54] Dobson, *Green Political Thought*, 161. For an excellent concise argument that a politically conservative stance necessarily implies an environmentalist one (focusing especially on this theme of prudence), see David W. Orr, "Conservation and Conservatism," in *The Nature of Design: Ecology, Culture, and Human Intervention* (New York: Oxford University Press, 2002), 97–103.

ponents of nature, and human beings with them, at serious and growing risk. And even in its own terms, Carson asserted, the drive to master nature through brute-force chemical technology was faltering, for "the insects are finding ways to circumvent our chemical attacks on them," rapidly evolving genetic resistance to the most commonly used compounds and profiting as well from pesticide-induced declines in some of their natural predators.[55] This account of the impending failure of the attempt to conquer insect pests exemplifies what Hirschman called the futility argument, and the prospect of actually having made them less manageable than before his perversity thesis. Aldo Leopold observed that human actions "have effects more comprehensive than is intended or foreseen," and he described an opposition between two ways of thinking, those of engineering and ecology. Engineers believed in reason and planning. They tended to assume "that a constructed mechanism is inherently preferable to a natural one" and to take direct and dangerously simplistic measures, using "crude and powerful" tools, to substitute the one for the other. That was engineering's theory; its practice led to many disappointments because of nature's unforeseen and indeed unforeseeable responses. The ecologist, who knew nature better, recognized that it was "too complex to enable him to predict its reactions."[56] Did not a writer like Hayek, though unsympathetic to environmentalism, precisely echo, in a different sphere, Leopold's criticism of "constructed mechanisms"? Is not "social engineering" a common target of conservative attack and Oakeshott's attack on "rationalism in politics" a rejection of the technical and engineering mentality in social affairs that environmentalists deplore when it is applied to biophysical ones?[57]

As Hirschman observed, progressives have not ignored the challenge posed by the unanticipated-consequences argument. The responses they have developed include, he noted, the claim that such consequences do

---

[55] Carson, *Silent Spring*, 245.

[56] Leopold, *A Sand County Almanac & Other Writings*, 183, 406, 410. Likewise, Barry Commoner, "A Cautionary Tale," in Joel Tickner, ed., *Precaution: Environmental Science and Preventive Public Policy* (Washington, DC: Island Press, 2003), 36, offered as classic environmental object lessons cases "in which unintended consequences of some new technology led to serious risks to the biosphere."

[57] Michael Oakeshott, "Rationalism in Politics," in Oakeshott, *Rationalism in Politics and Other Essays*, 1–36.

occur, but may well be trivial compared to the benefits of the intended change; that they occur and may at least as often be welcome surprises amplifying the benefits of the intended change as unwelcome ones detracting from it; and that over time our ability to foresee and deal with unexpected side effects increases and makes them less of a problem.[58] One can see examples of each of these kinds of arguments, which Hirschman ascribed to the left, in contemporary right-wing skepticism about the dangers of global climate change brought about by human emissions of greenhouse gases.[59] Some have argued that adapting to the challenges of climate change will cost much less than trying to prevent it, such that the change, though perhaps undesirable taken by itself, is not sufficient grounds to stop the activities that are bringing it about. Some have argued that a warmer world will be a better one for human beings, an unanticipated benefit of the combustion of fossil fuels. Some contend that the development of techniques for climate engineering, prompted in the first place by the challenges of adapting to inadvertent human-induced change, will, in the end, leave human society far better off than it was before, because they will furnish what never before existed, the knowledge needed to bring the natural processes that produce droughts and severe storms finally under control. Hayek, though acutely sensitive to the dangers of losing key elements of society at the simplifying hands of rationalistic planners, merely belittled the parallel environmentalist concern about the risks of fraying the web of nature: "were cockroaches to disappear," he wrote sarcastically, "the resulting ecological 'disaster' would perhaps not wreak permanent havoc on mankind."[60] Why do conservatives and progressives argue in opposite ways depending on whether the subject is society or the environment? If, as Hirschman maintained, such arguments reflect the deepest worldviews of each side, surely it would be

[58] Hirschman, *The Rhetoric of Reaction*, Chap. 6, "From Reactionary to Progressive Rhetoric," 149–163.

[59] As noted by Dryzek, *The Politics of the Earth*, 69. The first of these arguments is the one most commonly made by opponents of measures to prevent global climate change. For the second, see, e.g., Thomas Gale Moore, *Climate of Fear: Why We Shouldn't Worry About Global Warming* (Washington, DC: Cato Institute, 1998). Hamilton, *Earthmasters*, 107–109, 200–201, has discussed some examples of the third.

[60] Hayek, *The Fatal Conceit*, 27.

more consistent and more to be expected for the same arguments, and not the contrary ones, to appear as well in the two sides' earthviews.

Setting that puzzle aside for the moment, let us examine a second possible criterion for distinguishing left and right. Some authors, without disputing the reality and importance of the traditionalist/progressive opposition, have argued nonetheless that the principal basis of the left/right distinction lies elsewhere, in contrasting attitudes toward inequality within human societies. At the right-hand end, one finds a high degree of tolerance for such inequalities and correspondingly little receptivity to measures meant to reduce them. Any degree of inequality, conversely, becomes less and less acceptable as one moves toward the left-hand end. The criterion covers political, economic, and sociocultural questions alike, from the extent of the franchise to the distribution or redistribution of wealth to gender-based differences in opportunity to the exclusion or inclusion of marginal minorities. The Italian political scientist Norberto Bobbio wrote that "when we say that the left has a greater tendency to reduce inequalities, we do not mean that it intends to eliminate all inequalities, or that the right wishes to preserve them all, but simply that the former is more egalitarian and the latter more inegalitarian."[61]

The argument has a good deal of merit, and in practice, much more often than not, the labels of right and left employed alike by scholars and the general public indeed match such attitudes. Traditionalist and neoliberal conservatives have little difficulty in agreeing on this matter. Traditionalists see substantial inequalities as being in the necessary nature of things, as affirmed by the past experience of the human race. Neoliberals see them as necessary consequences of liberty and fear that heavy-handed measures to impose equality through redistribution will endanger liberty and the benefits, not merely economic, that it confers. Yet, the traditionalist right, at least, stands for something other than a mere vacuous and undifferentiated acceptance of inequality in general, or of more of it than progressives can tolerate, something that only its attitude toward tradition and innovation can explain. Conservatives feel

[61] For example, Norberto Bobbio, *Left and Right: The Significance of a Political Distinction*, trans. Allan Cameron (Chicago, IL: University of Chicago Press, 1996), 65. Noël and Thérien, *Left and Right in Global Politics*, also argued for this criterion as the central basis of the left/right division.

a particular tolerance for the kinds of inequality endorsed by tradition. Ones lacking its sanction they may very well dislike. If the distinction were as simple as Bobbio and others have stated it, it is not easy to see how there could be such a thing as right-wing populism with its strong charge of egalitarianism, though historians and political scientists have abundantly documented the existence in a variety of times and places of something that can only be called by that name. Nor, on Bobbio's premises, would the prominent position of the "liberal elite" in the demonology of today's American right be easy to understand. By invoking the role of tradition, though, one can readily account for it: the (supposed) power of the elite in question is illegitimate because it has no long and legitimating tradition behind it. Traditionalist conservatives may not be disturbed by the survival of some inequalities between men and women; they would surely balk at novel ones giving women greater power than men, just as they can without gross inconsistency criticize affirmative action programs as unfair and inequitable. Rather than being the sole basis of the left/right distinction, degrees of tolerance for inequality are a prevalent but not universal correlate of it. One might even try to subsume them under the criterion already examined, that of tradition or the status quo versus change. Those who are relatively well-off under the established order are likely to be its defenders, and those who are not are likely to criticize it for its inequities, such that attitudes toward tradition and change will closely correspond to ones toward equality.

Yet, this criterion itself, in its turn, usefully supplements that of progress versus tradition by adding more substance to it. Though we can specify the changes that conservatives oppose more precisely as those that break with tradition, it makes little sense to say that progressives are indiscriminately in favor of novelty, still less of change in general. Some novelties they would welcome, of course, but some they would abhor. Which ones? They would welcome those that represent progress, but what is that? First and foremost, it is what promises, in their eyes, to diminish inequality and increase the chances in life of the less well-off. The philosophy of John Rawls, which no one would hesitate about assigning to the left, is a classic expression of this basis for judging a change.

We might best, then, merge the two criteria discussed so far into a composite definition, as the American political scientist Ronald Inglehart did when he made it the test of left/right identification "whether one supports or opposes social change in an egalitarian direction."[62] (The left, of course, supports it.) But when we turn to environmental matters, instead of finding the answer to our earlier puzzle, we find instead a second one. Inglehart correctly spoke only of "social change" (excluding environmental change, that is), but why should he have had to? Why should the attitudes of left and right toward change of other kinds not follow the same lines of cleavage? And on the matter of equality, too, environmental attitudes do not line up with political-economic ones. The most straightforward way to reduce inequality is to make the poor richer, and the most straightforward way of doing that is to increase the abundance available, to make the earth more productive rather than to urge living within limits. Yet, it is a way that the right today advocates much more stridently than the left. It will not do to reply that the left merely doubts that Promethean environmental changes, as such, *would* be egalitarian in their effects. It perfectly well might expect them to be if they were undertaken by a progressive and egalitarian society, but that is not a hope consistent with environmentalism. And by the crudest criterion of all, seeing the right merely as the ideology of the haves and the left as that of the have-nots, we would expect concerns about long-term environmental sustainability or about the psychological benefits of natural experience to arise more readily among the haves, whose more basic needs have already been met, than among the have-nots, who have more immediately pressing things on their minds. Nor, finally, are nature's own workings—to put it mildly—conspicuous for their egalitarianism, which accounts for much of the left's longstanding suspicion of arguments that invoke nature as a model.

It is and long has been a characteristically right-wing maneuver to appeal to "nature" to justify the necessity and legitimacy of the existing order against those who would alter it. The rhetorical disconnect here between the worldviews and the earthviews of both left and right is so glaring that I need not belabor it. Bobbio was surely correct historically

[62] Ronald Inglehart, *Culture Shift in Advanced Industrial Society* (Princeton, NJ: Princeton University Press, 1990), 293. He added: "To speak in terms of Left and Right is always an oversimplification—but an extremely useful one."

when he claimed that "[t]he right is more willing to accept the natural and that second nature constituted by custom, tradition and force of the past," that it more often asserts the existence of "natural inequalities which cannot be attributed to society."[63] The modes of thinking commonly labeled Social Darwinism are but one example among many. The difference appears even in the imagery of debate, organic analogies for social processes, it is often noted, being more common on the right than on the left, and mechanical ones vice versa. In the *Reflections*, Burke not only observed that, in decrying inequality, "you are at war with nature," but also criticized the French Revolutionary constitution-makers for, "like their ornamental gardeners, forcing everything into an exact level," a comment at once on their egalitarianism and on their artificiality in contrast to the more unquestioned social hierarchies of English society and more natural tastes of English landscaping.[64] Burke's great critic Thomas Paine, as a biographer has observed, was in his writings fond of using analogies drawn from engineering works that altered and improved nature, especially from bridge-building.[65]

Avowed conservatives writing about another important and highly contentious realm of nature-society interactions, that of biotechnology, seem to feel quite comfortable invoking such principles as a human duty to respect nature, the ethical costs of abandoning traditional ways in the headlong pursuit of much-wanted goals that technology now brings within reach, and the need to be wary of the unforeseen consequences of meddling too deeply with things as they are as essentially conservative ones, in a way that this time seems quite akin to the conservative respect for human tradition. Leon Kass, later chairman of the White House Council on Bioethics under the second President Bush, approvingly quoted Hans Jonas at some length on the dangers of replacing the slow and proven workings of nature with hasty human choices that have not similarly been "exposed to the long test of the ages." Unless human beings choose "to learn our place in the natural whole and discover something of

[63] Bobbio, *Left and Right*, 67–68.

[64] Edmund Burke, *Revolutionary Writings*, ed. Iain Hampshire-Monk (Cambridge: Cambridge University Press, 2014), 50, 176.

[65] Edward Larkin, *Thomas Paine and the Literature of Revolution* (New York, NY: Cambridge University Press, 2005), 114–148.

its distinctive beauty and mysterious ground," he continued, "the project for mastery and possession of nature is a Faustian bargain." His fellow conservative bioethicist Yuval Levin likewise criticized "utopianisms of various stripes, all grounded in the dream of overcoming nature," and he invoked the restraints that a regard for stability, continuity, and traditional values should place on such a crusade.[66] Both, it is necessary to remember, were conservatives talking about biotechnology. Otherwise, looking only at the words I have quoted, one might easily mistake them for passionate greens assailing a reckless Prometheanism.

But perhaps here we have merely a confusion arising from that notoriously equivocal word "nature," one that can easily be resolved by untangling the different meanings of it in play. Andrew Dobson has made just that argument. The apparent contradictions, he suggested, do not really exist, because the two sides are not talking about nature in the same sense. The right respects chiefly the internal nature of the human being, the left, the world of external nature.[67] The observation is accurate, but it is not an explanation. What is it about conservatives, we still need to ask, that *should* make them more reverent toward the elements of nature within the human organism and progressives those outside it? It is all the more pressing a question in that the very distinction between the two realms is far from being a given that one simply must accept as part of the brute structure of reality. Writers such as the philosopher John Dewey and the physician and biochemist René Dubos have plausibly argued that the evolution, structure, and functioning of the human body are so necessarily and unavoidably correlated with surrounding conditions that one cannot usefully draw a line separating nature within and nature without. "In reality," Dubos wrote, "the internal environment should not be considered apart from the external environment." "The processes of living are enacted by the environment as truly as by the organism," in Dewey's words, "for they *are* an integration." The respiratory system makes no sense when it is treated as a purely internal matter, for it presupposes an external source of oxygen mixed in a certain ratio with other

---

[66] Leon R. Kass, *Toward a More Natural Science: Biology and Human Affairs* (New York, NY: The Free Press, 1985), 38, 153; Yuval Levin, *Imagining the Future: Science and American Democracy* (New York, NY: Encounter Books, 2008), 12, 28.

[67] Dobson, *Green Political Thought*, 163–165.

gases, and so on.[68] On such premises as these, a respect for one natural realm should logically go along, rather than conflict, with concern for the other. (Moreover, by the principal Western tradition of thinking on such matters, Judeo-Christian theology, human nature is fallen and corrupted, and external nature is not necessarily so. Hence conservative Christians, at least, might reasonably regard the latter as less in need of remaking than the former. And if their fallen state makes human beings unfit to meddle with their own nature, it might equally disqualify them from meddling with the rest of creation.) The puzzle remains.

One might, as Dobson also did, try to ground conservatism in a supposed belief in the unalterability of human nature, rather than in the undesirability of altering it.[69] Contemporary conservative anti-environmentalists might then claim that there are elements of human nature that stubbornly and inexorably demand the mastery and reshaping of the environment and are not to be denied. But it would not be a very different assertion that there are elements of human nature that are and always will be dissatisfied with human beings' own biological limitations and imperfections and will demand their remaking in spite of all that can be urged to the contrary. Yet, that is a claim that conservatives such as Kass and Levin do not seem to countenance. Why? And one would also, on such premises, be hard pressed to explain why Kass and Levin deplored the possible engineering of human characteristics precisely because they feared that it may become possible. They objected to it not as futile but as dangerous. Environmentalists object to the project of altering external nature on precisely the same grounds. Again the alignments we observe are not easy to account for in ideologically consistent terms.

Another defining difference sometimes asserted between left and right is the generalizing, absolutist, and universalist outlook typical of the former versus the particularist one of the latter. It can be derived

---

[68] John Dewey, *Logic: The Theory of Inquiry*, in *John Dewey: The Later Works, 1925–1953*, vol. 12, ed. Jo Ann Boydston (Carbondale and Edwardsville, IL: Southern Illinois University Press, 1986), 32; René J. Dubos, *Mirage of Health: Utopias, Progress, and Biological Change* (New York, NY: Harper, 1959), 102.

[69] Dobson, *Green Political Thought*, 163–165; Carter, *The Politics of the Environment*, 64, made a similar suggestion.

easily enough from the tradition/progress criterion, with the right more concerned to maintain traditions, in all their actual diversity, than the left, but is worth examining for a moment on its own. It accords, again, with the experience of the French Revolution, which as it developed became ever more notable for the strident universalism of its claims and aspirations, as did the political Marxism of the twentieth century. We tend, not without reason, to associate the universalism of the eighteenth-century Enlightenment with the left and the particularistic Romantic reaction of the early nineteenth century with the right. It is true, though, on the other hand, that right-wing thinkers have often been foremost in approving, and right-wing regimes in carrying out, the subjugation or suppression of minority traditions, against protests from the left. Yet, the protests, more often than not, have invoked universal human rights (stated, in this case, as a right to choose one's way of life), thus exemplifying rather than undercutting the distinction. Moreover, principled conservatives are not only, so to speak, particularists in the abstract; they are, at the same time, members—and, if they are consistent, loyal members—of some specific traditions, one of whose tenets may well be and often is its own objective superiority as a way of life over all the rest. Thus, in practice they may have to act in a way that in principle they may seem obliged to deplore. But of most interest here, again, are the environmental corollaries that may be drawn from these contrasting positions, and, again, we come up with a puzzle rather than a solution. The importance of preserving biodiversity and ecodiversity against the homogenizing pressures of modern technological civilization is a hallmark of environmentalist thought. Would not such rhetoric, again, come with more consistency from the right than from the left? Is it not, for example, echoed in the arguments of American, British, and French cultural conservatives against unrestricted immigration? Yet, the historian Peter Coates has found that, for all the apparent parallelism between the two discourses, in practice, in the modern USA they do not, in fact, go together. Those who appeal in debates on human immigration to the importance of maintaining

American cultural traditions tend to disregard or even denigrate environmental concerns.[70]

The most important candidate basis for the left/right split that I have not yet discussed is the alleged statism of the left and anti-statism of the right. What most fundamentally characterizes the right, many would say, is its dislike and distrust of government. If the claim were better grounded, our problem would be solved: conservatives eschew environmentalism because of the degree of government regulation that it seems to call for. But anti-statism is at once a pervasive element of Western, especially Anglo-American, conservative discourse today and quite useless for distinguishing left from right in earlier times. It furnishes the most immediately effective key to the puzzle of how contemporary cultural and libertarian or free-market conservatives, so-called, have been able to find enough common ground to hold together as the American Republican Party or the British Conservative Party or the various parties of the French rightist alliance.[71] Yet, the identification of conservatism with anti-statism is neither logically necessary nor, if one looks back through history, empirically dependable. Should we fly in the face of universal usage, as it would oblige us to do, by insisting that the leading anarchists of nineteenth-century Europe, such as Mikhail Bakunin and Enrico Malatesta, were really right-wingers because they disliked and distrusted the centralized nation-state, and should we call such conservative autocrats as Metternich and Nicholas I of Russia left-wingers because they did not? The questions answer themselves. Hayek regarded a "fondness for authority," which is hardly a synonym for anti-statism, as an essential element of conservative (and equally of socialist) thought.[72] Using the French Revolution, again, as a paradigm case gives us the same result. As

[70] Peter A. Coates, *American Perceptions of Immigrant and Invasive Species: Strangers on the Land* (Berkeley, CA: University of California Press, 2006), 151–164. Such linkages by some right-wing parties in several European countries, on the other hand, form the principal exception to the right/left environmental pattern there, but such parties focus only on environmental issues with which such connections can readily be made: Jonathan Olsen, *Nature and Nationalism: Right-Wing Ecology and the Politics of Identity in Contemporary Germany* (New York, NY: St. Martin's Press, 1999) and Damir Skenderovic, *The Radical Right in Switzerland: Continuity and Change, 1945–2000* (New York, NY: Berghahn Books, 2009), 206–214.

[71] Noël and Thérien, *Left and Right in Global Politics*, 89–95.

[72] Hayek, "Why I Am Not a Conservative," 522.

Tocqueville argued, the Revolution neither created nor reversed the transfer of authority in France to the central government; the victory of the left consolidated and furthered what the old regime had begun.[73] Their differences, which were indeed considerable, lay elsewhere. The greater sympathy of the right than the left for things military is anything but an expression of aversion to the state and especially the centralized nation-state. As late as the 1960s, few things more united the American right than a disdain for the Supreme Court under Earl Warren, but the Warren Court's most controversial decisions limited, rather than expanded, the powers of government—chiefly powers to do things that conservatives wanted it to do, such as easily convict criminal suspects, maintain racially separate school systems, conduct prayer in public schools, censor books and films, and restrict the distribution of contraceptives.[74] A consistently anti-statist right would have applauded much of the Court's work.

Anti-statism, then, is not a useful criterion for identifying the political right across time and space. It is an accidental and time-specific, though nonetheless an extremely important, characteristic of present-day Western conservatism. The state, for most people, is not a contested end but a means. Left and right, at different times, have welcomed or feared its activity chiefly according to whether it seemed likely to promote the particular ends that they have sought. Conservatism has been anti-statist not on principle, but when and where it seemed that the power of the state would be used to undermine traditions and hierarchies that conservatives prized, and not when it could be used to resist change and maintain tradition. As the suffrage has expanded and the Western liberal welfare state has emerged, conservatives have increasingly feared and denigrated a government whose energy they would have embraced and applauded if it had been deployed to different purposes. Hayek put the

---

[73] Jon Elster, ed., *Alexis de Tocqueville: The Ancien Regime and the French Revolution*, trans. Arthur Goldhammer (New York, NY: Cambridge University Press, 2011).

[74] For details, see Lucas A. Powe, Jr., *The Warren Court and American Politics* (Cambridge, MA: The Belknap Press of Harvard University Press, 2000). To be sure, the Warren Court also (though it aroused less controversy thereby) favored claims of government power to regulate the economy, for example in upholding the 1964 Civil Rights Act and numerous federal antitrust prosecutions. But that is the point. The factor common to the decisions that angered the Warren Court's conservative critics was not their statism, but their classically leftist egalitarianism and scant regard for tradition (illustrated also in the Court's rulings on legislative reapportionment).

matter squarely: because "conservatives are inclined to use the powers of government to prevent change" when they are able to, he observed, "the conservative opposition to government is not a matter of principle," only one of temporary expediency under certain conditions (he did not, it will be recalled, consider himself a conservative):

> the conservative does not object to coercion or arbitrary power so long as it is used for what he regards as the right purposes. He believes that if government is in the hands of decent men, it ought not to be too much restrained by rigid rules. … Like the socialist, he is less concerned with how the powers of government should be limited than with that of who wields them; and, like the socialist, he regards himself as entitled to force the value he holds on other people.[75]

The uselessness of this criterion for grounding the left/right distinction is more apparent with traditionalist or cultural conservatives than with neoliberal ones, and, indeed, Hayek considered the left-right spectrum less meaningful than a triangle, its three points conservatism, liberalism (in the minimal-government European sense), and socialism. But even the claims of Hayekian classical liberals to be principled anti-statists do not go unchallenged. Radical critics, beginning at least with Karl Polanyi, have argued that the anti-government rhetoric of market fundamentalists is much more show than substance. It conceals, they have asserted, a profound dependence on the state, not only to provide necessary services that the market relies upon but does not and cannot itself produce, but historically even to create what we think of today as a free market, to compel people to enter it, and subsequently to maintain it, by means including the use of state power to enforce private contracts; to create, distribute, and enforce private property rights; to create, maintain, and manipulate a monetary currency; to assess taxes that must be paid in cash, thereby forcing citizens into the commercial economy; and to step in and resolve dangerous economic crises that markets inevitably produce but cannot resolve.[76] One such critic argues that, in reality, rather than

[75] Hayek, "Why I Am Not a Conservative," 522–23, 525.

[76] Karl Polanyi, *The Great Transformation* (New York, NY: Farrar and Rinehart, 1944); Fred Block and Margaret R. Somers, *The Power of Market Fundamentalism: Karl Polanyi's Critique* (Cambridge,

opposing government action as such, conservatives support certain forms of it, notably "regulatory structures that cause income to flow upward," whereas "liberals support regulatory structures that promote equality … In the U.S. economy, there is no free market. It is just that structures that heavily regulate the economy are taken as inevitable."[77]

A similar but broader attempt to ground the distinction (and the politics of environmentalism) in a conflict between the claims of freedom or liberty defended by the right and those of regulation and control espoused by the left fails on even a cursory look. The distinction has very often been the opposite, as Hayek attested, the right upholding laws or customs curtailing personal freedom and its opponents urging their abolition. The associations of the phrase "law and order" certainly do not cluster on the left, nor those of "liberation" on the right. Progressives have often justified their economic egalitarianism on the grounds that poverty can be as important, or more so, in depriving people of freedom and in putting them under the thumb of others as can oppression by formal laws and regulations. And were a devotion to liberty in the broad sense—the freedom of human actions from control by some superior external authority—the defining criterion of conservatism, one would have trouble explaining what seems to be the stricter government within the family favored by self-described conservatives, usually in scornful opposition to the "permissiveness" of more progressive-minded parents. But if we understand conservatism as an attachment to tradition, the difficulty disappears. Conservatives certainly see hierarchy and order in the family as imperatives endorsed by the weight of longstanding precept and practice, a consideration that for them trumps any countervailing concerns about individual freedom.

Nor can we say that the essential difference between progressives and conservatives is that the former care more about human well-being than the latter (or vice versa). They merely disagree on how much well-being is possible and on how it is best obtained and safeguarded. Here, the three arguments that Hirschman identified come frequently into play. What progressives see as beneficial reforms, conservatives are apt to see

MA: Harvard University Press, 2014).

[77] Dean Baker, *Taking Economics Seriously* (Cambridge, MA: MIT Press, 2010), 1–2.

as futile and unrealistic attempts to remake the necessary order of things, or as likelier to cause harm than good. Though progressives often regard such arguments as merely rationalizing the privileges that some enjoy from the status quo, so, too, conservatives suspect that ambitious programs for social betterment conceal the selfish motives of those who hope to ride them into power more than they do a genuine concern for the good of humanity. If environmentalists argue that the degradation of the environment jeopardizes human well-being, Prometheans respond that what jeopardizes it is rather the curbs on technological and economic development, the most plausible means of advancing well-being, that environmentalists advocate. They indeed often accuse environmentalists of sacrificing people's welfare—and no less that of future generations than of the present—to the interests of nature. We are no closer than before to understanding why human-induced change, viewed from the left, appears so much more of a danger to human beings than it seems from the opposite side.

But perhaps environmentalism is more common on the left than on the right because the left's favored discourse of "rights"—early exemplified in Thomas Paine's *The Rights of Man*, his defense of the French Revolution against Burke's *Reflections*—has led by logical development and extension to the notion of rights of nature, whereas conservatism's hostility to the idea of such general principles that override the dictates of tradition and authority has prevented any such development on its side.[78] Even if true, though, and it may well be, it would not answer the question at issue here. It could, at best, explain the political affinities of an ecocentric environmentalism that asserts the independent claims and intrinsic value of other forms of life. But most environmental concern is not ecocentric. It derives, rather, from concerns about human well-being. Hans Jonas grounded his argument for prudence and restraint in the engineering of the environment squarely on the imperative of ensuring the survival of the human species. If one's concern is for nature itself, he wrote, one need not strive to avert catastrophes, for that is a term of evaluation and in nature there is no such thing, no change being less natural

[78] The possibility is explored in Roderick Nash, *The Rights of Nature: A History of Environmental Ethics* (Madison, WI: University of Wisconsin Press, 1989).

than any other; catastrophes exist only by human standards of reckoning.[79] Opponents of environmentalism have caricatured Carson's *Silent Spring* as expressing more worry about birds and bugs than about people. Yet, in fact it put vastly more weight on the threat that the reckless use of synthetic pesticides posed to human health, to long-term agricultural production, and to elements of the environment (such as songbirds) that human beings value as part of the quality of their own lives than it did on any essentially ecocentric concerns. Stratospheric ozone depletion threatened both people and other species, but it was dealt with as promptly and effectively as it was only because of the dangers to the former. The prospect of global climate change would arouse no more than a small fraction of the fears it has evoked if it were thought a menace only to other species and not to the survival of human life and civilization. Left and right may indeed differ in their sensitivity to the well-being of the rest of nature, but their disagreement on that point cannot explain the gap in their other and most salient environmental concerns. It has little to say about why the former see certain environmental changes as far more dangerous to human survival than the latter do.

But if views of ethics will not explain the matter, perhaps attitudes toward science will. Our problem would be solved if we could but ascribe the differing receptivity of left and right to what science tells us about the dangers of climate change and of environmental change in general to a characteristically greater respect on the left for its authority. But though this might, if true, explain the left's present-day environmentalism, it would not explain the anti-environmentalism of the contemporary right. Why should the latter have abandoned all of its other reasons for skepticism about radical Promethean change merely for the sake of disagreeing with whatever science says? Nor, in fact, can one say that the left is inherently or in practice more obedient to the authority of science than the right is. The differences between the two basic ideologies do not imply any such contrast. One might argue that they do, that science, being progressive, rational, universal, and empirical, necessarily unsettles social tradition and local particularism and with them the social orders that they underpin. It played just that role in the arguments and rhetoric of

[79] Jonas, *The Imperative of Responsibility*, 188.

the eighteenth-century Enlightenment. Yet, one could equally well argue that it must necessarily sustain the given social order by furnishing the image of a given, necessary, and law-governed natural order, unchangeable by human volition. Science studies what is, and it seeks to explain why what happens, happens—thereby implying that it had to happen—and thus always justifies the status quo as necessary against claims that things could be different (hence the appeal to many conservatives, though not to all, of forms of biological and environmental determinism and, as noted earlier, of imagery drawn from the nature that the natural sciences study). These and other conflicting possibilities cancel each other out. The fact that both right and left have been portrayed as anti-science in principle or practice, each by comparison with the other and with a wealth of examples on both sides, strongly suggests that neither is so. Each, rather, seems to be favorable or hostile to the apparent implications of specific scientific research findings, the right seizing upon those that seem to support the value of tradition and the inevitability of inequality, the left the opposite. Neither displays comprehensive support or hostility for science as a whole; neither offers many examples of scientific findings inducing its adherents to change any politically inspired beliefs they previously held.[80] We are left again with the puzzle of why the right has received scientific warnings in the realms of climate and the environment so much more skeptically than the left.

A rather desperate possibility remains to be considered: that change—particularly technological change, population growth, and economic growth, with all the environmental transformations that they produce—has become so common that it is now itself a tradition, and as such legitimately dear to conservative hearts. In the form in which I have stated it, the argument looks paradoxical, and the appearance is no illusion, for it amounts to saying that if innovation only becomes routine enough it wins the support, as a kind of meta-tradition, of those who prize tradition and

[80] Two books that appeared in the same recent year claimed, each citing numerous cases in point, that certain elements in the ideologies of conservatism and progressivism, respectively, lead to a distrust of science and its findings: Chris Mooney, *The Republican Brain: The Science of Why They Deny Science and Reality* (Hoboken, NJ: John Wiley, 2012) and Alex B. Berezow and Hank Campbell, *Science Left Behind: Feel-Good Fallacies and the Rise of the Anti-Scientific Left* (New York, NY: PublicAffairs, 2012). For additional considerations on both sides, see also Levin, *Imagining the Future*, Chaps. 5, "Science and the Left," and 6, "Science and the Right."

distrust innovation. It seems, in any case, too much an ad hoc hypothesis invented to overcome a local difficulty, for when we test the same line of reasoning elsewhere it fails. Surely there is just such a "tradition of change" in modern times in family and sexual mores, for example. Do traditionalist conservatives, as the argument implies they would, cheer on its further progress? No; quite the opposite, and they have mobilized, too, to resist the nineteenth- and twentieth-century trend toward an increasing scope for government in Western countries. The argument also fails entirely to account for conservatives' greater suspicion of biotechnology than of environmental engineering. A weaker form is less implausible but still inadequate: that conservatism, with its respect for current ways of living, resists environmentalism's claim that they are leading to disaster and must be restrained and, indeed, reversed. For "turning back the clock" is not ordinarily a concept disagreeable to the right; most of the ways of living in question are new, not traditional, having only shallow roots in human history; and when the assertion is made that they are on a collision course with the natural foundations of all possible human existence, it would surely be the more consistently conservative impulse to give priority to preserving the latter by restraining the former. Finally, it would be no less paradoxical, and no more convincing, to argue that, because conservatism is a worldview of limits and restraint, it entails limits and restraint on restraint—say, on environmental regulation—itself. Conservatives who genuinely valued the preservation of nature would regard moderation in its defense, to paraphrase Barry Goldwater, as no virtue and extremism on its behalf no vice.

To recapitulate, two criteria, singly or in combination, offer the most plausible grounds for the distinction between the political left and right: attitudes toward tradition and toward egalitarianism. It matters little which, if either, of these criteria one adopts as the principal basis for the left/right distinction, for they can be examined separately, as well as in combination. Today they show consistent associations with environmentalism and Prometheanism in environmental matters, but associations the opposite of those that one might expect. If this discussion so far has failed to find a convincing reason why they do, it has at least made a case for a second and related project. If today's pattern of environmental politics is, after all, consistent with the deepest beliefs defining the opposed worldviews, it should have existed in much the same form in the past as it does in the present. We might better understand it if we expanded the scope of the inquiry and asked whether it held true in earlier times as well. If it did, there would be all the more reason to keep looking for the elusive factors that necessarily associate Prometheanism with the right, environmentalism with the left. If it did not, we might do better to seek more historically contingent and accidental causes for today's alignments.

What ideological affiliations characterized the politics of environmental Prometheanism's major past (pre-1960) exponents? Were these thinkers generally identifiable with the left or the right, and did their views, more specifically, fall on the traditionalist or progressive, hierarchical or egalitarian, and, for what it is worth, the statist or the anti-statist sides of the debates of their day? Such a review would do much to clarify how deep or shallow are the roots of contemporary ideological cleavages on the environment.

Anyone who is still skeptical of the whole enterprise, who even at this point in the argument thinks today's pattern too inevitable even to admit the possibility of its reversal, might consider a figure from the past, one in whom the three themes of progressivism, human transformation of the environment, and the myth of Prometheus come together, the English poet Percy Bysshe Shelley (1792–1822). Shelley belonged to the first generation born into a Europe reconfigured by the French Revolution, and his political sympathies lay entirely on the left. He disbelieved in Christianity and other traditional religious creeds and was expelled from Oxford for refusing to disavow authorship of a pamphlet entitled *The Necessity of Atheism*. He advocated disestablishment of the state church and the extinction of other legacies from the dark ages of what he called "error and fanaticism."[81] He equally rejected such established political institutions as the monarchy and a parliament dominated by an aristocratic class. A convinced pacifist, he insisted on the necessity of nonviolent change, while partly excusing the excesses of the Terror in France as the unfortunate result of the corrupting effects on the populace

[81] James Bieri, *Percy Bysshe Shelley: A Biography* (Baltimore, MD: Johns Hopkins University Press, 2008), 114–126; Percy Bysshe Shelley, *A Philosophical View of Reform*, ed. T.W. Rolleston (London: Oxford University Press, 1920), 13.

of long-continued despotism. Strongly egalitarian in his economic views, he deplored the "unequal distribution" of the produce of labor brought about by, and further reinforcing, the unequal power of the rich and the poor. He advocated the payment by the rich of the national debt and looked forward as an ultimate ideal to an "equality of possessions" among mankind.[82] Fully agreeing with his father-in-law, the radical philosopher William Godwin, he harshly criticized the Malthusian theory of population as a libel on the poor, whom, both thought, it falsely depicted as responsible for their own poverty. (Godwin himself noted with disapproval that "[t]hree fourths of the habitable globe are now uncultivated" and argued that even the area already farmed was capable of being made much more productive.[83])

By any of the usual defining standards, then, Shelley stood on the far left of the political debates of his time. He combined his politics with a visionary environmental Prometheanism that associated humankind's liberation from political despotisms of rank and tradition with its reformation of nature. As a student at Oxford, more absorbed in scientific experiments than in his coursework, he speculated enthusiastically to a friend on how the discovery of a cheap and easy way of manufacturing water might alter the earth's surface for the better: "The arid deserts of Africa may then be refreshed by a copious supply and may be transformed at once into rich meadows and vast fields of maize and rice." With a similar command over the production of heat, he continued, one might warm "even the most ungenial climates as readily as we now raise the temperature of our apartments to whatever degree we may deem agreeable or salutary." Chemistry might likewise provide the means of equalizing "the remarkable fertility of some lands, and the hopeless sterility of others."[84] A few years later, in an odd anticipation of Goethe's Faust, Shelley threw himself and his money enthusiastically behind a large engineering project

[82] *Ibid.*, 16–17, 50, 56–60, 70. See also Paul Foot, *Red Shelley* (London: Sidgwick and Jackson, 1980).

[83] Mayhew, *Malthus*, 94–97; William Godwin, *An Enquiry Concerning Political Justice and Its Influence on Modern Morals and Happiness*, ed. Isaac Kramnick (New York, NY: Penguin, 1985), 769.

[84] Thomas Jefferson Hogg, *Shelley at Oxford*, ed. R.A. Streatfield (London: Methuen & Co., 1904), 17–19.

to reclaim land on the coast of Wales from what he called "the unfruitful sea" for settlement and cultivation (a project that ended in almost as much discord as the German's fictional one).[85] He praised the republican institutions of the USA for fostering the "cultivation and improvement of the soil" in a way that the high concentration of land ownership in England tended to frustrate. He thought to confute believers in a wise God, be they orthodox or deist, by pointing out how many features in "the economy of the globe" in its existing state were, for human purposes, ill-designed. With a bow to Bacon, he hailed the progress of invention by which the land had been "compelled to furnish more and more subsistence," while regretting that an evil system of society had failed to turn the increase to most people's benefit.[86]

In his poetry, Shelley offered lyrical glimpses of the earth remade once society had been remade. His most ambitious early long poem, *Queen Mab* (1816), is at once a political and an environmental utopia. As society is reformed, the poem maintains, so too will the globe be: the deserts watered and made fertile and prosperous, the ice of the polar regions melted, the tempests of land and sea quieted. It envisions, once all of the nobler human qualities become predominant, "each unfettered o'er the earth extend/Their all-subduing energies, and wield/The scepter of a vast dominion there." For now, though, the poem sadly concludes, "a pathless wilderness remains/Yet unsubdued by man's redeeming hand."[87] *Prometheus Unbound* (1819–1820) closes with an ecstatic vision of humankind "[c]ompelling the elements with adamantine stress."[88]

Today it is difficult not to think the two strains in Shelley quite incongruous. Some recent critics who seem to find his politics congenial have found his environmental Prometheanism deeply disturbing. One described it, in light of present-day concerns, as "a tragically misguided

[85] Timothy Morton, *Shelley and the Revolution in Taste: The Body and the Natural World* (New York, NY: Cambridge University Press, 1994), 229–231 (quotation from 230).

[86] Shelley, *A Philosophical View of Reform*, 11, 13; Percy Bysshe Shelley, "A Refutation of Deism" (orig. 1814), in E.B. Murray, ed. *The Prose Works of Percy Bysshe Shelley*, vo. 1 (Oxford: Clarendon Press, 1993), 109.

[87] Donald H. Reiman and Neil Freistat, eds., *The Complete Poetry of Percy Bysshe Shelley*, v. 2 (Baltimore, MD: Johns Hopkins University Press, 2004), 224–225, 229, 234.

[88] M.H. Abrams, ed., *The Norton Anthology of English Literature*, 4th ed., vol. 2 (New York, NY: W.W. Norton, 1979), 731.

vision."[89] On the other hand, Ron Arnold of the late twentieth-century right-wing American Wise Use movement praised *Prometheus Unbound* for its forthright message of global environmental engineering, as "a single, tight expression of the wise use sense of life," though his politics and Shelley's have little in common.[90]

But if, by today's standards, Shelley seems a deeply confused thinker, the confusion may lie in our minds and not in his. Some other possibilities deserve attention before we dismiss him as simply muddleheaded. Perhaps he was an isolated anomaly, rational enough in his thinking, but in a highly idiosyncratic way. Or perhaps the answer Dryzek has proposed is correct: that environmental Prometheanism was so dominant an earthview in the nineteenth-century West that it infected thinkers of the left and right impartially. Finally, it might be that today's affinities were reversed, that environmental Prometheanism before the mid-twentieth century was more characteristic of left than of right, and that today's allegiances would strike a thoughtful visitor from the past as no less bizarre than Shelley's seem to us. By attempting and failing above to explain the logic of the present-day pattern, I have already made a case for this last possibility as the one that some ways we might expect to find: a *progressive Prometheanism*, making little sense within the present-day discourse of environmental politics, from which it is all but absent, but which may once have seemed a necessary corollary of leftist views on political and social matters.

Prometheanism before about 1960 is, as I have noted already, not quite terra incognita to historians of environmental thought, but still little-known and poorly mapped territory. I begin, therefore, with the most modest of the tasks that arise from the possibilities I mention above: determining whether Shelley's mix of progressivism and Prometheanism was unusual or even unique. To find that it was neither would, though

[89] Eric Gidal, "'O Happy Earth! O Reality of Heaven!': Melancholy and Utopia in Romantic Climatology," *Journal for Early Modern Cultural Studies*, vol. 8, #4 (2008), 74–101 (quotation from 75); see also P. M. S. Dawson, "'The Empire of Man': Shelley and Ecology," in Betty T. Bennett and Stuart Curran, eds., *Shelley: Poet and Legislator of the World* (Baltimore, MD: Johns Hopkins University Press, 1996), 232–239, and Michael R. Page, *The Literary Imagination From Erasmus Darwin to H. G. Wells: Science, Evolution, and Ecology* (Farnham, Surrey: Ashgate Publishing Limited, 2012), 67.

[90] Arnold, "Overcoming Ideology," 23.

not disproving Dryzek's suggested answer, still be of considerable interest, for it would mean that leftist or progressive political views are not at all incompatible with environmental Prometheanism. It such were the case, we could probably abandon the search for some timeless logic to the alignment of earthviews and worldviews we see today and look instead for more historically specific factors to explain it.

Did many notable individuals or schools of thought champion a Promethean remaking of the earth from the end of the eighteenth century until about 1960 and associate it with a progressive politics? I look mainly for thinkers or movements of recognized stature, the better to determine whether, to some of the best minds of their day, political progressivism and environmental Prometheanism seemed compatible or not. Each, to be included, must have elaborated a case in general and overarching terms for the desirability of human transformation and improvement of the environment (a Promethean earthview), and must have connected it to a clearly stated progressive worldview. A global inventory, of course, would be desirable, but I confine my examination to four countries: the USA, Great Britain, France, and Russia.[91]

And my answer, to state it in advance, is that there was nothing odd in its time or for a long time afterwards in Shelley's combination of worldview and earthview. I provide my evidence in the next three chapters, which are organized not in strict chronological order or by country, but with a chapter devoted to each of three cohorts, cutting across the boundaries of period and nation: those whom I call technocrats, natural scientists, and prophets (the divisions, as will be apparent, are not sharp ones), to illustrate the varieties as well as the commonalities of past progressive Prometheanism. My findings, then, return us to the question that arises next and that I take up in the conclusion: was enthusiastic environmental Prometheanism merely a part of the common presuppositions of the age, or was it more commonly associated with the left than with the right, the reverse of what we see now?

[91] I can read sources in the original in English, French, and Russian. Another reason for choosing these countries is their objective importance for the topic at hand, the centrality of three of them to modern Western thought and the usefully contrasting marginality of the fourth.

2

The Technocratic Prometheans: Engineering Society and Environment

The Green parties that began to emerge in the 1980s, most prominently in Germany, were not what they may have seemed to be, the first political movements in the Western world to make environmental questions an important part of their platform. Yet, it is not difficult to understand why their chief precursors have been overlooked. The earliest, Saint-Simonianism in early nineteenth-century France, also had much to say about the effects of human activity on the earth's surface. Unlike the Green parties of later years, though, it took an almost wholly positive view of them. It combined an enthusiastic environmental Prometheanism and glorification of technology, particular Green *bêtes noires*, so to speak, with political views closely akin to those prevalent in late twentieth- and early twenty-first-century ecologism: progressive, egalitarian, pacifist, and hostile to the unrestrained hegemony of the market. Manifestoes from its heyday in the early 1830s mingled appeals for equality of the sexes and for the abolition of poverty with denunciations of war, armaments, and militarism and with fervent calls to "exploit, fertilize, beautify the globe on which we live, smooth down the mountains, drain the marshes, confine the rivers and the sea within conquering dikes, furrow the ground

W.B. Meyer, *The Progressive Environmental Prometheans*,
DOI 10.1007/978-3-319-29263-2_2

with canals and railroads … to appropriate matter in a thousand different ways for the ever-growing needs of humankind" and to intensify "the beautification and exploitation of the terrestrial globe."[1] One might well ask whether one movement or the other did not simply err in regarding its own portfolio of beliefs as consistent and coherent. Certainly few people today seem to find any discordance among the elements that make up the Green credo. What of the earlier one?

Even as a notably ungovernable young man, Henri, comte de Saint-Simon seemed guided by impulse in certain directions more than others. Born in 1760 into an old family of the French aristocracy, he crossed the Atlantic in 1779 to fight for the independence of the American colonies from the British crown. The war concluded, he drifted south to Mexico to present—unsuccessfully—a plan to the Spanish viceroy for uniting the Atlantic and Pacific oceans by a water route across the Central American isthmus. Returning to Europe, he was offering to help the government of Spain revive a stalled project for constructing a navigable waterway between Madrid and the ocean when the outbreak in France itself of another revolution, with which he initially sympathized, called more loudly for his services.[2] He was drawn, in short, to projects for improving both nature and society. The end of the Terror found Saint-Simon still alive and, though chastened, recognizably the same man. He now set himself to devising a new social order that would be capable, as existing ones seemed not to be, of mastering the world of nature for the good of all, and he slowly began to attract a coterie of talented disciples. Robbed of his presence by his death in 1825, they and an array of newer recruits, a disproportionate number of them graduates of the elite state engineering academy, the *École Polytechnique*, were at the same time liberated to develop his doctrines, which they did with great energy and creativity.[3]

[1] *L'Organisateur: Journal de la doctrine saint-simonienne* 2, #21 (8 January 1831), 164; 2; #24 (29 January 1831), 192.

[2] Frank Manuel, *The New World of Henri Saint-Simon* (Cambridge, MA: Harvard University Press, 1956).

[3] Antoine Picon, *Les saint-simoniens: Raison, imaginaire et utopie* (Paris: Belin, 2002), 102–112; Pamela M. Pilbeam, *Saint-Simonians in Nineteenth-Century France: From Free Love to Algeria* (Houndmills: Palgrave Macmillan, 2014), 12.

Saint-Simon and his followers emphasized the paramount importance for human well-being of "industry," perhaps the single most important word in their lexicon. Science, in Saint-Simonian usage, meant knowing; industry meant doing. It was industry that had applied scientific knowledge to the work of mastering nature. It turned sandy wastes and marshes into fertile plains, controlled and redirected the flow of rivers, flattened mountains, altered and improved other species, and generally had begun the "new and progressive evolution that man and the planet on which he lives must one day undergo. Such is *industry*, and those who carry out these works are the *industrials*."[4] In the past, both the earth and human society had progressed, Saint-Simon observed, but only slowly and through terrifyingly violent catastrophes that punctuated periods of long stagnation. The application of reason could tame the process in both realms and make it a much more orderly one, as well as a more rapid and efficient one.[5] But that could only be done if society's institutions were refashioned to correspond to its new needs. Existing institutions were sadly outmoded and ineffective, relics of a vanished or vanishing era that had been dominated by conquest and force and been justified by a revealed supernatural theology. They had little to offer a new age whose chief concern must be the generation of wealth and the betterment of humankind by peaceful industry. Politics Saint-Simon defined as "the science of production," or of creating the proper social framework for the maximization of production, which he defined in turn as the successful exploitation of the resources of the earth.[6] The seats of political authority, still held by the last representatives of the era of conquest, military power, and theological doctrine, must be filled instead by a different set of leaders—on whom society already, de facto, most depended—better equipped to meet the challenges of the present.

An incident that occurred in the early years of the Bourbon Restoration dramatized the point too vividly for comfort. In the fall of 1819, Saint-Simon asked readers to imagine what would happen to France if the

[4] *Oeuvres de Saint-Simon et d'Enfantin* (Paris: Dentu/E. Leroux, 1865–1878) (hereafter, *OSSE*), vol. 42, 403–404; vol. 44, 23–24.

[5] *OSSE*, vol. 44, 24–25.

[6] *OSSE*, vol. 18, 188–90.

individuals who held its most honored positions—members of the royal dynasty and of the hereditary nobility, ministers of state, cardinals and bishops of the Church, and generals of the army—were, by some sad chance, all to vanish from the earth. It would be personally most regrettable, he observed, but it would be no catastrophe for France. The lower ranks of society contained many individuals more than capable of filling the vacancies. But imagine, he continued, France's most knowledgeable and creative minds in the pure and applied sciences and arts being similarly swept away. They could not be replaced nearly so easily, and their loss would mean the country's decline and impoverishment. The little parable, as Saint-Simon called it, was pointed enough, but to the government it became downright sinister in February of 1820 when a radical fanatic assassinated the heir presumptive to the throne, the duc de Berry. Saint-Simon faced charges of fomenting revolution. After lengthy proceedings, a jury acquitted him, but the episode had stamped his ideas, and not inaccurately, with a subversive character.[7]

Progressive in its small regard for tradition and its high hopes for the future, the Saint-Simonian movement was progressive, too, in its egalitarianism. Its aim, made ever more explicit after 1820, was the reorganization of society in the interest first and foremost of—in a phrase of Saint-Simon's that his disciples made into a kind of mantra after his death—"the most numerous and the poorest classes" of the population. It sought to equalize the conditions and opportunities enjoyed by all, raising the general standard of living through the enhancement of society's capacity to produce, while dissolving established hierarchies of rank and power (including, to a degree unusual for the time, those between men and women).[8] It was pacifist in seeing war and conquest as irrational and outmoded wastes of resources in an industrial age. It was statist in its emphasis on the need for conscious, authoritative direction of the whole of society's efforts and resources towards its goals.

None of these are characteristics one particularly associates with twenty-first century environmental Prometheanism, yet Saint-Simonianism was

[7] Manuel, *The New World of Henri Saint-Simon*, 210–214.

[8] Pilbeam, *Saint-Simonians*, 25–40; Claire G. Moses, "Saint-Simonian Men/Saint-Simonian Women: The Transformation of Feminist Thought in 1830s France," *Journal of Modern History* 54, #2 (1982), 240–267.

resolutely Promethean as well. Its leaders saw industry and engineering as nothing if not the application of tools and human intellect toward the mastery of the physical world. Viewing, as they did, the intelligent, planned subjugation and exploitation of nature as society's principal task, they drew the lesson that technicians—the foremost "industrials," the engineers or applied scientists and those who directed their work—were, as the experts in such matters, the people to whom the task should be entrusted. The most visible and flamboyant of Saint-Simon's disciples, Prosper Enfantin (1796–1864), faulted the society of his time for chronically undervaluing the human faculty of "constructivity" in favor of many more elegant but trifling forms of intellect, an error common, he wrote, among those who "do not grasp what it is to manipulate a world and sculpt it in the image of God." A genius for constructivity marked the great inventors and engineers, he continued, ones "who have never written a book, written a line of metaphysics, of philosophy, of history, of physiology, of literature," yet "who have pierced, surmounted, crushed mountains, filled valleys, traversed rivers."[9] Saint-Simon himself characteristically admired the beaver above all other animals. Naturalists erred, he wrote, in thinking that apes were the species closest in intelligence to human beings. Beavers possessed two traits that he particularly prized: the ability to work collectively and the itch to refashion their surroundings.[10]

Saint-Simonianism epitomized the technocratic variety of environmental Prometheanism, one that emphasizes Prometheus's gift, that of fire, and the tools and technologies that it symbolizes. It highlighted, as modern Prometheans do, what Enfantin called "constructivity" and the twentieth-century American historian James C. Malin dubbed "the contriving brain and the skillful hand."[11] Technocratic Prometheans look to human ingenuity to furnish the necessary means for reshaping everything that is unsatisfactory or substandard in physical geography, exalting the plains, lowering the mountains, making the rough places plain and the

---

[9] *OSSE*, vol. 46, 209–210.

[10] *OSSE*, vol. 40, 49–52, 109; see also vol. 38, 182–183.

[11] James C. Malin, *The Contriving Brain and the Skillful Hand in the United States* (Lawrence, KS: the author, 1955). Malin adopted the phrase from the nineteenth-century American city promoter Jesup W. Scott.

deserts and the swamplands blossom, and squeezing far more out of the earth's resource base than it was capable of providing in its natural state.

Saint-Simon expressed a profound regard for Francis Bacon and for René Descartes, European philosophy's early prophets of environmental Prometheanism. In his writings, most vividly in the short utopian fiction *The New Atlantis*, Bacon had seen the control of the forces and riches of nature, "the enlarging of the bounds of Human Empire," as society's principal task and the source of ever-increasing wealth, comfort, and happiness for its members. Descartes had similarly justified the pursuit of scientific inquiry by its promise "to make ourselves, as it were, the lords and masters of nature."[12] But both Bacon and Descartes had more to say about the methods of scientific research that would forge the tools of such mastery than about the political and social conditions that might foster or hinder their application. Writing in the twentieth century, Michael Oakeshott characterized Saint-Simon's work as the most important attempt in European thought to extend the logic of Bacon's ideas to the political and economic spheres. How, it asked, must human society be organized in order to ensure the development and proper use of the technologies that enable it to master nature? Like Bacon, but more clearly Oakeshott, wrote, Saint-Simon "perceived that this entailed rulers who are technologists," possessing a definite vision of the tasks to be done, and a social system giving them the authority to direct people's actions accordingly. Saint-Simonianism represented, in Oakeshott's eyes, the first statement of the case for a form of politics that would become a powerful force in the later nineteenth and twentieth centuries, a productivist collectivism that would culminate in the ideology of the Soviet Union.[13] For Friedrich Hayek, too, the Saint-Simonian movement represented a pioneering attempt, as misguided in his view as it was influential, to apply

---

[12] *The Philosophical Works of Francis Bacon*, ed. James Spedding, vol. 3 (London: Longmans, 1861), 156; René Descartes, *Discourse on Method and Related Writings*, trans. Desmond M. Clarke (London: Penguin, 1999), 44.

[13] Michael Oakeshott, *Morality and Politics in Modern Europe: The Harvard Lectures*, ed. Shirley Robin Letwin (New Haven, CT: Yale University Press, 1993), 100–107 (quotation on 105).

the methods of the natural and engineering sciences to rationalizing society for the sake of human betterment.[14]

In politics, a technocratic vision can, as it did with Saint-Simon, harmonize readily with a disdain for tradition and for conservatism. In an age of rapid advances in a society's capacities for production, nothing hampers the adoption of new and better and more efficient ways of doing things more than an attachment to the past and a suspicion of the "newfangled." In this spirit, the technocrat will agree with Bacon that the supposed wisdom of the past is a poor and meager thing compared to the accumulated knowledge and skills of the present, which the future in turn will far surpass. If age and experience bring wisdom, Bacon proposed, antiquity was the callow adolescence of the race and not its venerable oracle. The golden age, Saint-Simon asserted in the same spirit, lay in the future and not, as tradition had it, in the past.[15]

These radical, anti-traditionalist overtones are not the only ones that a technocratic politics can possess. Under some circumstances, the proposal to put power in the hands of technicians may rather have a conservative character. Social and political conflicts and the wishes of the marginal and the disenfranchised may then tend to be ignored in favor of a focus on the efficient operation of the institutions that underpin the status quo. Technicians may then be those who least question the legitimacy of things as they are. The label "technocratic" is, indeed, sometimes used disparagingly to denote an approach that ignores inequalities and the political questions and conflicts to which they give rise. It so happens, at any rate, that the principal proponents of Western technocracy in the nineteenth and early twentieth centuries coupled it with a hostility to tradition and also with an ideal of social progress achieved through the alteration of the natural world.

For nothing, likewise, hampers the adoption of new and better ways of arranging the earth's surface than a reverence for the established, natural

[14] F.A. Hayek, *Studies on the Abuse & Decline of Reason: Texts and Documents*, ed. Bruce Caldwell, vol. XIII in *The Collected Works of F. A. Hayek* (Chicago, IL: University of Chicago Press, 2010), Part II, "The Counter-Revolution of Science," 169–281.

[15] James Spedding, Robert Leslie Ellis, and Douglas Denon Heath, eds., *The Works of Francis Bacon*, vol. 4 (London: Longman & Co., 1858), 82; M. le comte de Saint-Simon and Augustin Thierry, *De la réorganisation de la société européenne* (Paris: Adrien Egron, 1814), 111–112.

order of things. The radical German engineer John A. Roebling came to the USA from Germany in 1831, when Saint-Simonianism was at its peak in Europe, seeking both a progressive land of freedom and equality and an opportunity for applying his skills to "employ nature's forces and make them our slaves," to transform the Sahara into a garden and do the same with the American wilderness, altering even "its climate and seasons."[16] To those of this cast of mind, natural history is analogous to human history, and each is a story of progress, lamentably slow when left to its own devices but capable of being accelerated by rational planning. To resist change in either is to hold onto inferior modes of organization inherited from a less proficient and rational past, which, even if they might once have worked decently well, can no longer do so under the new conditions of the present.

After the French revolution of 1830 toppled the restored Bourbon dynasty, the Saint-Simonians offered the new and more liberal king, Louis-Philippe, the advice that what the country needed was work and not fighting. They presented a list of projects that could be undertaken immediately: the improvement of the Paris water system, the systematic clearing for intensive cultivation of the wildlands of the Vendée region, the digging of new canals connecting with the Loire River, a network of railroads across all of France. Vast public works, they suggested, useful in themselves, would also offer a useful outlet for the labor freed by mechanization. Beyond the short term, they envisioned a similar remaking of the rest of the world, including the piercing by canals of the two great isthmuses that blocked major world shipping routes, Suez and Panama.[17] Exuberant young contributors expounded these and other projects over the next couple of years in the newly Saint-Simonized Paris daily *Le Globe* and elsewhere. Widely read in France and beyond, they caught the attention of, among others, the aged and increasingly conservative Goethe. Literary historians have interpreted the diabolical project of environmental engineering in the second half of *Faust* as his response.[18]

[16] Quoted in Alan Trachtenberg, *Brooklyn Bridge: Fact and Symbol* (Chicago, IL: University of Chicago Press, 1965), 40.

[17] *OSSE*, vol. 6, 156–158.

[18] See, e.g., Hans-Jürgen Schings, "Magicians of Modernity: Cagliostro and Saint-Simon in Goethe's *Faust II*," in *Goethe's Faust: Theatre of Modernity*, ed. Hans Schulte, John Noyes, and Pia

The *Globe*'s editor, the young Saint-Simonian Michel Chevalier (1806–1879), was a characteristic representative of the movement. He had a fittingly wide-ranging eye for one whose goal was the conquest of the entire earth. He wrote enviously to an American sympathizer of the challenges facing the inhabitants of the New World, their most fertile lands still waterlogged and their mightiest rivers the sources of devastating floods. The things that they were doing to conquer nature, he wrote, formed a model for others: covering wildernesses with harvests, irrigating fields, draining marshes, diking rivers, linking lakes and streams by canals and making them pathways for steamboats as the lands were becoming for railroads.[19] Closer to home, Chevalier drew up a blueprint for a "Mediterranean system" of improvements that would reawaken the lands of classical civilization, North Africa and continental Europe as far north as Germany and Russia. It would stitch them together in a net of railroads, canals, dredged rivers, improved ports, and a canal through the isthmus of Suez, and it would facilitate the productive exploitation of all of their agricultural, mineral, and human resources, irrigation and drainage canals making now-desolate soils bloom. In time, Chevalier added, a Panama Canal would promote similar links with and within the Americas.[20] Noting what vast obstacles the earth's terrain posed to many of these projects, he deplored how little had been done to advance the technology of blasting since the invention of gunpowder centuries before. Christian and humanitarian prejudices against creating dangerous and destructive substances, he suspected, had hindered such research, obscuring the realization of how much good could be done by their peaceful use to reshape the land surface.[21]

Definite and practical in some moods, as suited the trained *Polytechnicien* he was, Chevalier was equally capable of rhapsodies about the Saint-Simonian liberator who would bring the human race to an understanding of its tasks. This visionary hero would fill valleys and lower mountains, set up a beacon of enlightenment above the cities of

---

Kleber (Cambridge: Cambridge University Press, 2011), 78–93.

[19] *OSSE*, vol. 8, 201.

[20] Michel Chevalier, *Système de la Mediterranée* (Paris: Bureau de la Globe, 1832), 36–38, 47–48, 50, 51–53.

[21] Ibid., 5–12, 52n–53n.

the world, trace the routes of rapid communication between them in order to unite them in a single faith, free proletarians and women from their slavery. So common did this style become within the movement, that one of its leaders, Saint-Amand Bazard, just before leaving it, cried off on "these vast plans, these gigantic projects thought up that morning, matured during the day, and ready to print in the evening, in which, from one pole to the other, the races, the nations, the oceans, the rivers, the marshes, the deserts, the valleys, the mountains are united, brought closer, fertilized, cut, traversed, surmounted …."[22]

The more immediate cause of Bazard's secession was the increasingly radical doctrine of sexual liberation advocated by his erstwhile co-prophet, Enfantin.[23] Equally uneasy were the functionaries of the July Monarchy that had come to power in 1830. They proved little more open to Saint-Simonian advice on policy than the restored Bourbons had been, and they undertook to repress the movement in 1832 amid rumors of scandalous goings-on among the men and women of the sect at its country retreat at Ménilmontant, outside Paris. Enfantin and Chevalier, convicted of offenses against morals, received sentences of a year in prison each.[24] The crackdown had the effect of breaking up Saint-Simonianism as an organized sect, but likewise of spreading its influence by dispersing its leaders. Shortly after his release, Chevalier began a distinguished career in government service, culminating under the Second Empire of Napoleon III, which engaged the services of many onetime Saint-Simonians and pursued many of the school's aims. The man whom Bazard's departure had left as the unchallenged leader, Prosper Enfantin, eventually followed his former colleague's advice to forsake grand visions and concentrate on detailed and disciplined plans for getting things done. He took himself to Egypt and for several years devoted himself to the projects of a dam on the Nile and a ship canal through the Isthmus of Suez. He later moved to Algeria, which France had acquired as a colony in 1830, and

[22] *OSSE*, vol. 6, 182–184; vol. 7, 69.; see also his contribution to the "Livre Nouveau" assembled by the group in their Ménilmontant retreat: Philippe Régnier, ed., *Le Livre Nouveau des Saint-Simoniens* (Tusson, Charente: Du Lérot, 1991), 247–248.

[23] Sebastien Charléty, *Histoire du Saint-Simonisme (1825–1864)* (Paris: Paul Hartmann, 1931), 125–136; Picon, *Les saint-simoniens*, 131–138.

[24] Charléty, *Histoire du Saint-Simonisme*, 175–204.

busied himself with plans for settling and cultivating its arid wastelands, emphasizing in his work the advantages of collective and communal over fragmented individual action. Like Chevalier, he obtained a measure of official influence on the coming of the Second Empire.[25]

Enfantin's greatest achievement was, perhaps, to have interested a French consul in Egypt, Ferdinand de Lesseps (1805–1894), in the project of a Suez canal, which in time became the latter's consuming interest.[26] Not himself an engineer, Lesseps deployed remarkable talents as an organizer and manager in the period between the start of work in 1854 and the triumphal opening of the Canal in 1869. Feted by the world in general and France in particular as the greatest conqueror of nature's barriers who had ever lived, he allowed success and acclaim to go to his head. In 1879, he assumed charge of what seemed his logical next task, the excavation of a sea-level passage through Panama to unite the Atlantic and the Pacific.

While thus engaged, Lesseps also found time to champion another Promethean project, that of the French engineer François Élie Roudaire (1836–1885) for the creation of a vast interior sea in France's North African possessions of Tunisia and Algeria, where a large tract of arid wasteland lay below the level of the Mediterranean, separated from it only by the narrow Isthmus of Gabès. A canal dug through that barrier, Roudaire pointed out, would fill the depression with a large new body of water. Evaporation from its surface would moisten and cool the climate of the surrounding lands and make them a fertile garden, one opened to fructifying contact with the outside world, moreover, by the shipping and port cities that the new sea would accommodate. The scheme, which Roudaire pressed on the French government and public for years, was that of a committed socialist from his youth—not a Saint-Simonian, but a follower of the French utopian Charles Fourier (of whom more later), and an apostle of linked improvement in the social and the natural order.[27]

---

[25] Ibid., Books III and IV, 205–346; Picon, *Les saint-simoniens*, 153–164.

[26] Charléty, *Histoire du Saint-Simonisme*, 305–312; Picon, *Les saint-simoniens*, 240–242; Pilbeam, *Saint-Simonians*, 125–129.

[27] On the history of the project, see Jean-Louis Marçot, *Une mer au Sahara: Mirages de la colonisation, Algérie et Tunisie* (Paris: La Différence, 2003); on Roudaire's socialism, 206–219.

In the midst of his labors at Panama, Lesseps won election to the French Academy, where he took his seat on April 23, 1885. To a sitting member, the celebrated scholar of ancient Middle Eastern religions Ernest Renan, fell the task of making a speech of welcome in response to the newcomer's own inaugural address. Over the course of a long life, Renan's political sympathies wandered about a good deal, particularly his attitudes toward progress, democracy, and science. To most of the outside world, and particularly to traditionalistic, Catholic, conservative Frenchmen, though, they were nothing if not simple. Renan was the apostate who had left the Catholic priesthood as a young man; he had then devoted himself to scholarship that undermined the timeless truths of the Bible and the Church's traditions and gave comfort to atheists, Republicans, and socialists. In welcoming the conqueror of Suez, Renan indeed made a memorable statement of faith in the idea of earthly progress. He barbed his welcome with a few thorns; he recalled the Saint-Simonian roots of the project, which Lesseps, in his own speech, had just taken some pains to discount, and to his celebratory words he added the dark prediction that the Suez Canal, so vital and so strategically placed, would undoubtedly become a great battleground of the world's future wars. All the same, he spoke warmly and appreciatively of what the new Academician had done. To make itself fully at home on the world in which it lived, Renan averred, humankind had to correct the unfortunate configuration that nature had in many places given it. If we could survey the other planets of the universe, he suggested, we could judge their level of civilization by a simple test. Had they, or had they not, pierced the isthmuses that blocked their direct routes of ocean travel? By this test, the earth's inhabitants were just then crossing an important threshold. The nineteenth century, Renan observed, with Lesseps as its instrument, was cutting through the earth's two most inconveniently placed barriers, those of Suez and Panama. He reassured the man who had overseen the work that he need not fear the reproach of impiously tampering with the divine creation. He had clearly improved upon God's plan.[28]

---

[28] *Discours de réception de M. Ferdinand de Lesseps; Réponse de M. Ernest Renan* (Paris: Calmann Lévy, 1885), 6, 17–18, 20, 34, 39.

Part of the eulogy turned out to be premature. Lesseps failed as spectacularly at Panama as he had succeeded at Suez, his enterprise collapsing in the early 1890s in a nightmare of frustration, malaria, and yellow fever at the worksite and financial and political scandal in Paris. But it was not long before the Central American isthmus was pierced, though by a lock canal rather than the Frenchman's planned sea-level passage. The USA took up the work in 1903 and opened the completed waterway to traffic in 1914. The French effort had been a private enterprise, financed by investors who had hoped to match or exceed the returns earned by shareholders in the Suez Company. The American government managed the Panama Canal's construction and paid the bills. Some Americans at the time held up the contrast between the French failure and their own country's success as evidence of the superiority of state effort and its indispensability in the largest tasks in the conquest of nature.[29]

The contrast was an unusual one where the two countries were concerned. The USA, when they differed, has more typically opted for private enterprise and France for state direction. But the Americans, for whom Panama exemplified the superiority of government work undertaken by scientific experts over projects driven by the profit motive, were speaking from the wavecrest of one of Arthur M. Schlesinger's "tides of national politics," the Progressive era in the USA.[30] They and their compeers looked to the government, and especially the federal government, to rationalize and regulate the chaotic forces of society. An influential offshoot of Progressivism, the American conservation movement, had similar plans for the natural world. Its program, the historian Bruce Schulman has written, fused "governing nature" with "nurturing government," looking to state rather than private and market management of the earth and

[29] Willis J. Abbot, *Panama and the Canal in Picture and Prose* (New York, NY: Syndicate Publishing Company, 1913), 117–118, 169; Emory Adams Allen, *Our Canal in Panama* (Cincinnati, OH: United States Publishing Company, 1913), 194, 197–199, 232–237; Joseph Bucklin Bishop, *The Panama Gateway* (New York, NY: Charles Scribner's Sons, 1913), 58, 255, 258–272. On this theme, see also Alexander Missal, *Seaway to the Future: American Social Visions and the Construction of the Panama Canal* (Madison, WI: University of Wisconsin Press, 2008), 77–78.

[30] Arthur M. Schlesinger, "The Tides of National Politics," in *Paths to the Present* (New York, NY: Macmillan, 1949), 77–92, an idea rephrased and updated by Arthur M. Schlesinger, Jr., "The Cycles of American Politics," in *The Cycles of American History* (Boston, MA: Houghton Mifflin, 1986), 23–48.

its resources. The government, the conservationists believed, would both exploit the earth more efficiently (replacing the wastefulness of markets and competition with the expert knowledge of government scientists) and distribute the proceeds far more equitably. Their goal, described by Schulman as "a new American state … structured on the administrative management of natural resources," had much in common with that of the Saint-Simonians.[31]

Like the Saint-Simonians, the American conservationists were distinctly Promethean progressives. They meant by the word "conservation" something very different from what it means in common usage today, and they clashed frequently (though they also found common ground on occasion) with "preservationists," John Muir foremost among them, who wanted to protect nature as it was. (Preservationists, as a biographer of one of them has noted, often stood to the right of the conservationists on issues not involving the environment.)[32] The central figure in the progressive conservation movement was the forester-turned-politician Gifford Pinchot (1865–1946).[33] Pinchot had been trained in forestry as an applied science, the management of tree cover for the output of timber and other benefits that it yielded. His vision of political economy merged neatly with his vision of the environment. Each, he thought, fared best when managed by experts. Neither reached anything like its maximum potential for bettering the human condition when left to a chaotic and wasteful competitive free-for-all. Under individual ownership and distribution by the market, Pinchot wrote, not only had some resources been used wastefully, but "we have left unused vast resources which are capable of adding enormously to the wealth of the country."[34] "[T]he

[31] Bruce Schulman, "Governing Nature, Nurturing Government: Resource Management and the Development of the American State, 1900–1912," *Journal of Policy History* 17, #4 (2005), 375–403 (quotation from 382).

[32] Jonathan Peter Spiro, *Defending the Master Race: Conservation, Eugenics, and the Legacy of Madison Grant* (Hanover, NH: University Press of New England, 2009), 58. The classic works on the American conservation/preservation divide are Samuel P. Hays, *Conservation and the Gospel of Efficiency: The Progressive Conservation Movement, 1890–1920* (Cambridge, MA: Harvard University Press, 1959) and Nash, *Wilderness and the American Mind.*

[33] Char Miller, *Gifford Pinchot and the Making of American Environmentalism* (Washington, DC: Island Press, 2001).

[34] Gifford Pinchot, *The Conservation of Natural Resources*, United States Department of Agriculture Farmers' Bulletin #327 (Washington, DC: Government Printing Office, 1908), 9–10.

fundamental principle of the whole conservation policy," Pinchot stated, "is that of use, to take every part of the land and its resources and put it to that use in which it will best serve the most people." Under private enterprise, resource exploitation benefited the few far more than it did the many. Scientific state management would mean "a land subdued and controlled for the service of the people" and not for "the profit of the big man."[35] The late twentieth-century American "wise use" movement took its name from Pinchot's writings, and not altogether inappropriately. It shared his view of nature, though not his political and economic ideals.

The historian Samuel P. Hays memorably dubbed the core credo of progressive conservation "the gospel of efficiency." Its leading figures found nature's processes as wasteful as those of economic competition. Pinchot, in Donald Worster's words, "saw the world as badly in need of managing, and he was convinced that science could teach man to improve on nature, to make its processes more efficient and its crops more abundant."[36] On some points their thinking overlapped with those of the preservationists, but the latter rejected the Promethean element in conservation that sought frankly to improve on nature by controlling and reconstructing it. Pinchot "was less interested in preserving nature untouched than in standing guard to make sure it was used in the wisest, most efficient way."[37]

Both conservationists and preservationists favored measures for the maintenance of forests, but with fundamental differences. Conservationists such as Pinchot urged that tree cover be managed on a utilitarian basis. Attentive to such indirect services that it provided as the stabilization of streamflow, they had little interest in preserving or restoring natural scenery or forest communities. In a stance as calculated to appall modern environmentalists as his statism would appall modern Prometheans, Pinchot criticized New York State for its bad example in setting apart its

---

[35] *Hetch Hetchy Dam Site: Hearings Before the Committee on the Public Lands, House of Representatives, Sixty-Third Congress, First Session, on H. R. 6281* (Washington, DC: Government Printing Office, 1913), 25; Gifford Pinchot, *The Fight for Conservation* (New York, NY: Doubleday, Page & Company, 1910), 27, 29.

[36] Hays, *Conservation and the Gospel of Efficiency*; Donald Worster, *Nature's Economy: A History of Ecological Ideas*, 2nd edition (New York, NY: Cambridge University Press, 1985), 267.

[37] Ted Steinberg, *Down to Earth: Nature's Role in American History*, 3rd edition (New York, NY: Oxford University Press, 2013), 137.

Adirondack holdings to be preserved forever as "wild forest lands" not to be cut for timber.[38] The economist Richard T. Ely, another leading conservationist, pointed out how much more productive a managed forest was than a wild one.[39] Pinchot and his acolytes instituted policies for the indiscriminate suppression of forest fires, natural no less than human-caused, which they regarded as deplorably wasteful. They thought old-growth forests best logged for the sale of their timber, making way for vigorously productive younger trees. Conservationist managers of public lands sought to reshape their zoogeography as well, eradicating predator, "pest," and other undesired species and propagating the ones favored by hunters and sightseers.[40]

Pinchot also wanted to develop "every use to which our rivers can be put, and every means available for their control." The government geologist W J McGee, the movement's chief authority on water management as Pinchot was on forests, wrote that the "conquest over nature" would not be complete until it included "the conquest of the waters"; "at no distant day," he predicted, "the running waters of the earth will be wholly subjugated and sent hither and thither at man's behest." Conservationists proposed dams, reservoirs, levees, locks, and dredging along the nation's rivers to serve a variety of purposes simultaneously: power generation, irrigation, municipal water supply, flood control, and navigation. McGee became the prophet of multiple-purpose river development and especially of a "Lakes-to-Gulf Deep Waterway," a ship channel to be dredged and maintained between the Great Lakes and the Gulf of Mexico in association with an array of dams and other projects. Turning the entire Mississippi River system into a regulated canal, he acknowledged, would

---

[38] Pinchot, *The Fight for Conservation*, 55; Philip G. Terrie, *Forever Wild: A Cultural History of Wilderness in the Adirondack Park* (Syracuse, NY: Syracuse University Press, 1994), 106–107, 122. The conservationists' program for reforestation no more made them forest preservationists than people who advocated the demolition and new construction of a city neighborhood could be called architectural preservationists merely because they, too, proposed that the land be covered with buildings.

[39] Richard T. Ely, "Conservation and Economic Theory," in Richard T. Ely et al., *The Foundations of National Prosperity: Studies in the Conservation of Permanent National Resources* (New York, NY: Macmillan, 1917), 5.

[40] Steinberg, *Down to Earth*, 140–143; Nancy Langston, *Forest Dreams, Forest Nightmares: The Paradox of Old Growth in the Inland West* (Seattle, WA: University of Washington Press, 1995).

be a stupendously vast and difficult task, but one well worth the effort, not least for the example that it would set for other and more easily mastered streams elsewhere. It would place a natural feature presently "as lawless as a monster of the jungle" under human mastery, making the tamed waters cease to harm the settlements on their banks and instead furnish an abundance of benefits.[41] In their most famous conflict with Muir and other preservationists, the conservationists applauded the construction of a dam to fill the Hetch Hetchy Valley in California, part of the Yosemite National Park, with a reservoir. They emphasized the greater claims of human well-being (abundant water and electricity for the people of nearby San Francisco, under clauses guaranteeing public control of both utilities) over those of natural scenery. Pinchot himself presented these arguments in favor of the dam at a Congressional hearing.[42] A number of its supporters promised that, by creating a beautiful lake where none had existed, the dam would make the valley far more sightly than before, human management improving upon nature even in aesthetics.[43] Few things better symbolize what the progressive conservationists stood for than a large dam, just as few better symbolize what modern environmentalists oppose.

Nature's follies, in the conservationists' opinion, as well as letting rivers run wastefully wild in times of flood, included the chronic spoiling of potentially useful lands by excessive moisture, an error they saw manifested over a large swath of the USA. Wetland "reclamation," the large-scale drainage of swamps and marshes through aggressive government action, formed a central plank of the conservation program. Pinchot deplored the condition of "millions upon millions of acres, now lying waste in swamps all over the country, but capable of supporting

---

[41] Pinchot, *The Fight for Conservation*, 55; Hays, *Conservation and the Gospel of Efficiency*; W.J. McGee, "Water as a Resource," *Annals of the American Academy of Political and Social Science* 33, #3 (1909), 39; W.J. McGee, *The Earth the Home of Man*, Anthropological Society of Washington Special Paper 2 (Washington, DC, 1894), 26; W.J. McGee, "Our Great River," *World's Work* 18 (February, 1907), 8576–8584 (quotations from 8579 and 8580).

[42] Nash, *Wilderness and the American Mind*; Robert W. Righter, *The Battle over Hetch Hetchy: America's Most Controversial Dam and the Birth of Modern Environmentalism* (New York, NY: Oxford University Press, 2005).

[43] Righter, *The Battle over Hetch Hetchy*, 102–105, 121.

in comfort millions of people."[44] Writing in 1907, a Department of Agriculture drainage engineer and conservationist, James O. Wright, was more precise: the USA east of the Rockies held seventy-seven million acres of such watery waste, equivalent to the area of England, Ireland, Scotland, and Wales combined.[45] Wright explained why the government must take the lead through drainage laws to organize the work if this vast territory were to be conquered from nature. The uncoordinated efforts of individual landowners could not possibly transform extensive wetlands into healthful and productive acreage, and the costs of bargaining and the impossibility of forcing unwilling owners to do their part prevented coordinated private action. Only the state could initiate and direct the work, force would-be free riders to cooperate and contribute, and apportion the costs fairly among all those benefited. It could likewise provide the expert direction for want of which many private efforts had failed.[46] Wright shortly thereafter drew up the plans for a project to drain the entire Florida Everglades for cultivation and settlement, a pet cause of two notably progressive early twentieth-century governors of the state, William S. Jennings (served 1901–1905) and Napoleon Bonaparte Broward (served 1905–1909).[47]

Conservationists equally promoted the transformation by irrigation of lands that in their natural state seemed too arid to farm, a territory within the USA even vaster than the domain of undrained wetlands. The irrigation movement's leading evangelist, William E. Smythe, was at once Promethean and progressive. As Patricia Limerick wrote, Smythe celebrated a "human war against nature," a "war to transform wilderness into submissive farmland"; the desert "appeared in Smythe's world as an

[44] Pinchot, *The Conservation of Natural Resources*, 11.

[45] J.O. Wright, *Swamp and Overflowed Lands in the United States: Ownership and Reclamation*, United States Department of Agriculture Experiment Station Circular 76 (Washington, DC: Government Printing Office, 1907), 8.

[46] Ibid., 11–12, 13–14.

[47] Christopher Meindl, Derek H. Alderman, and Peter Waylen, "On the Importance of Environmental Claims-Making: The Role of James O. Wright in Promoting the Drainage of Florida's Everglades in the Early Twentieth Century," *Annals of the Association of American Geographers* 92, #4 (2002), 682–701; Michael Grunwald, *The Swamp: The Everglades, Florida, and the Politics of Paradise* (New York, NY: Simon & Schuster, 2006); David R. Colburn and Richard K. Scher, *Florida's Gubernatorial Politics in the Twentieth Century* (Tallahassee, FL: University Presses of Florida, 1980).

enemy," something to be overcome and metamorphosed by human will into something better. He coupled his enthusiasm for watering the deserts with a political and economic program that emphasized settlement in cooperative colonies and a heavy reliance on government action rather than competitive private enterprise.[48] He and other Western irrigation crusaders represented the task as too large to be carried out successfully other than through government, and particularly federal government, action. It required the scientific expertise available at the national level, the authority to coordinate interstate allocation of water, and the revenues generated by previous projects to finance new ones. Their program took shape in the National Reclamation Act (the Newlands Act) of 1902 and the establishment of a federal Bureau of Reclamation. The Bureau's first director, Frederick Haynes Newell, an MIT-trained engineer, was another quintessential progressive Promethean. Newell disparaged the ability of private enterprise to reclaim the West and emphasized the conquest of aridity by government agencies and experts for the sake of human welfare. He fought vigorously, though unsuccessfully, to have a national program of wetland drainage added to his duties. Reclamation of both kinds he defined as "substituting the will of man for the unregulated natural forces."[49]

The breakaway Progressive Party that Theodore Roosevelt led in the election of 1912 upheld against both Republicans and Democrats a program of "extending, instead of limiting, the powers of government" in the interest of human well-being.[50] Its adherents included the principal conservationists and wrote into its platform a large array of environmental measures, almost without exception Promethean proposals for the exploitation or transformation of nature. Calling for the conservation of

[48] William E. Smythe, *The Conquest of Arid America* (New York, NY: Macmillan, 1905) and *Constructive Democracy: The Economics of a Square Deal* (New York, NY: Macmillan, 1905); Patricia Limerick, *Desert Passages: Encounters with the American Desert* (Albuquerque, NM: University of New Mexico Press, 1985), 84, 88.

[49] Donald C. Jackson, "Engineering in the Progressive Era: A New Look at Frederick Haynes Newell and the U.S. Reclamation Service," *Technology and Culture* 34, #3 (1993), 539–574; Newell quoted in Anthony E. Carlson, "The Other Kind of Reclamation: Wetlands Drainage and National Policy, 1902–1912," *Agricultural History* 84, #4 (2010), 457.

[50] Theodore Roosevelt, *Progressive Principles: Selections from Addresses Made during the Presidential Campaign of 1912*, ed. E.H. Youngman (New York, NY: Progressive National Service, 1913), 212.

America's natural resources, the drafters explained what they meant by conservation: control and use by the nation rather than monopolization by private owners. These resources, they declared, "must be promptly developed and generously used to support the people's needs." Roads should be extended and improved and a system of national highways created. "The coal and other natural resources of Alaska should be opened to development at once." Waterways should be improved and managed for navigation, power generation, and flood control. The Mississippi River's floods should be brought under control and the Lakes-to-Gulf deep waterway constructed as a federal project. Cultivable land should be reclaimed from deserts and wetlands "to support millions of people" and the public grazing lands more intensively managed to increase their productivity.[51]

The two longtime members of Congress who most authoritatively represented the Progressive spirit were Robert M. La Follette, senator from Wisconsin from 1906 until his death in 1925 and the presidential nominee of a second Progressive Party in 1924, and George W. Norris, congressman and then senator from Nebraska from 1903 to 1943. La Follette voted for the Hetch Hetchy dam, and *La Follette's Weekly Magazine*, his political organ, made it clear in 1912 how a Progressive ought to understand the term "conservation," which it defined succinctly as "Wise Use, Not Foolish Disuse":

> True conservation consists not in hoarding our resources, but in using them properly. Our waterpowers running night and day from year to year without turning a wheel are of no value to the public. To permit the mature trees of our forests to rot in waste is not conservation. To deny to this generation the advantage of the proper development of our coal fields and other mineral wealth, is to deny to them participation in the benefits which rightly belong to them. The problem before us is not to hoard our resources but to develop them in such a way that the benefits from development will inure not to a few men, but to the rightful owners—all the people of the United States.[52]

[51] Donald Bruce Johnson, ed., *National Party Platforms, vol. 1: 1840–1956* (Urbana, IL: University of Illinois Press, 1978), 179–180.

[52] *La Follette's Magazine* 4, #7 (17 February 1912), 3–4.

The 1924 Progressive Party that nominated La Follette for President, like its predecessor of 1912, gave an important place in its campaign literature to environmental—but not environmentalist—measures. It strongly endorsed the Lakes-to-Gulf deep waterway and the public control and aggressive development of America's key natural resources. It called for full exploitation of the nation's waterpower potential "now running away down the river beds," the control of rivers to prevent floods, the irrigation of "millions of acres of arid lands," and the public production for sale at cost of synthetic fertilizers.[53]

"Conservation to Norris," his biographer wrote, likewise "did not involve locking up natural resources; rather, it provided for their wisest and fullest use according to the most advanced scientific knowledge and techniques of scientific management."[54] Blaming scarcity and high cost, as did most Progressives, on the profit motive, the Nebraskan in the 1920s suggested that the government own and manage the most important natural resource in the country, the coal mines, to make fuel more affordable to the public. He voted for the Hetch Hetchy reservoir and spoke at length in the Senate against its preservationist critics. He saw chiefly the sinister hand of private utility interests, rather than any valid concerns of preservationists, behind much of the opposition to public river development, at Hetch Hetchy and elsewhere.[55] Throughout his career, he promoted the impoundment and development of rivers for the multiple uses of irrigation, flood control, navigation, power generation, and synthetic fertilizer production, always by public agencies rather than private capital. Along with California's Hiram Johnson, Roosevelt's Progressive running mate in 1912, Norris was the Senate's most ardent champion of the Boulder Dam to bridle and harness the Colorado River, and he fought less successfully for decades for a dam and reservoir on the Potomac River to supply water and electricity to the nation's capital.[56]

---

[53] Johnson, *National Party Platforms*, 253, 254–255; *The Facts: La Follette-Wheeler Campaign Text-Book* (Chicago, IL: La Follette-Wheeler Campaign, 1924), 42, 66, 67–68, 69, 113–114.

[54] Richard Lowitt, *George W. Norris: The Persistence of a Progressive, 1913–1933* (Urbana, IL: University of Illinois Press, 1971), 23.

[55] Ibid., 22–24, 191; Righter, *The Battle over Hetch Hetchy*, 130.

[56] Lowitt, *Norris*, 25–27, 191–193, 264–267, 351–354; Richard Coke Lower, *A Bloc of One: The Political Career of Hiram W. Johnson* (Stanford, CA: Stanford University Press, 1993), 231–237.

He greeted the creation in 1933 of the Tennessee Valley Authority, his greatest legislative achievement, as "the dawning of that day when every rippling stream that flows down the mountain side and winds its way through the meadows to the sea shall be harnessed and made to work for the welfare and comfort of man." Toward that end, he pushed for similar authorities on other American rivers and overseas as far afield as the Danube and the Jordan.[57] On Norris's death in 1944, David Lilienthal, one of TVA's founding directors, wrote:

> Just a day or two before he was fatally stricken, Senator Norris received word that must have meant a great deal to him. The word was that TVA had just closed the great steel gates on the Kentucky dam, and had begun to impound water behind this, the last and largest dam to be built on the main stem of the river. Thus for the first time in history man had placed a great river completely in his control and forced it to work in the interest of humankind, a memorial, for the centuries, to the insight and love for his fellow-man of a wise, simple, and courageous American.[58]

The words characterize their author as much as they do their subject. Lilienthal fully shared Norris's enthusiasm for the construction of dams and locks to master rivers, restrain their floods, and exploit their potential for power and navigation. He left TVA after the war to promote the peaceful use of atomic energy under President Truman and subsequently the construction of TVA-type river projects in the Third World.[59] Truman himself combined the progressive liberalism of his "Fair Deal" with a parallel technocratic activism in environmental matters. He launched a "Point Four" program to export the means of resource exploitation to the

[57] Alfred Lief, *Democracy's Norris: The Biography of a Lonely Crusader* (New York, NY: Stackpole Sons, 1939), 420; George W. Norris, "A TVA on the Jordan," *The Nation* 158, #2 (20 May 1944), 589–591. On the association between political progressivism and the control of nature in this period, see also Wesley Arden Dick, "When Dams Weren't Damned: The Public Power Crusade and Visions of the Good Life in the Pacific Northwest in the 1930s," *Environmental Review* 13, #3/4 (1989), 113–153 and Ronald A. Foresta, *The Land Between the Lakes: A Geography of the Forgotten Future* (Knoxville, TN: University of Tennessee Press, 2013), Chaps. 1 and 2.

[58] David Lilienthal, "Senator Norris and the TVA," *The Nation* 159, #13 (23 September 1944), 343–344.

[59] Steven M. Neuse, *David Lilienthal: The Journey of an American Liberal* (Knoxville, TN: University of Tennessee Press, 1996).

world's poorer countries, one about which conservative Republicans were as skeptical as they were about TVA. In meetings with Lilienthal in 1949, he "waxed enthusiastic" about TVA-type programs in the Middle East, India, Africa, Brazil, and China, wondering only whether atomic energy might soon make hydroelectric power dams obsolete. "We might even use an atomic bomb to change the course of some river," Lilienthal recorded him as saying.[60] Truman compiled a similarly Promethean record in the domestic sphere, keeping up the New Deal's emphasis on large dams, irrigation, and land and resource development.[61] It was an outlook he shared, their other differences notwithstanding, with the third Progressive Party revolt, that of 1948, which bolted his nomination from the left. The Progressive platform criticized both Republicans and Democrats for abandoning "the American dream of abundance." In terms that might as well have come from Saint-Simon and his disciples, it identified "the limitless potential of modern technology" and "the planned development of all of our resources" as the twin keys to realizing that dream. It called specifically for "the peaceful application of atomic energy," a particular enthusiasm of the party's presidential nominee, Henry A. Wallace, just as the public control of large rivers for power generation, flood control, and irrigation, which the platform also celebrated, was a particular enthusiasm of Wallace's running mate, Senator Glen H. Taylor of Idaho.[62]

Progressivism as an organized political movement lasted longest west of the Mississippi. It was legislators from that part of the country who most durably combined radical politics with enthusiasm for great

---

[60] David Ekbladh, *The Great American Mission: Modernization and the Construction of an American World Order* (Princeton, NJ: Princeton University Press, 2010) (on Republican opposition to Point Four as to TVA, 100–101, 153, 163, 234); Thomas Robertson, "Conservation after World War II: The Truman Administration, Foreign Aid, and the 'Greatest Good,'" in Karl Boyd Brooks, ed., *The Environmental Legacy of Harry S. Truman* (Kirksville, MO: Truman State University Press, 2009), 32–47; *The Journals of David E. Lilienthal, vol. 2: The Atomic Energy Years, 1945–1950* (New York, NY: Harper & Row, 1964), 525–526, 593–594.

[61] Karl Boyd Brooks, "Introduction: Los Alamos to the Everglades—Harry S. Truman's Environmental Legacy," in Brooks, *The Environmental Legacy of Harry S. Truman*, xvii–xxxi.

[62] Johnson, *National Party Platforms*, 438, 443; Mark L. Kleinman, *A World of Hope, a World of Fear: Henry A. Wallace, Reinhold Niebuhr, and American Liberalism* (Columbus, OH: Ohio State University Press, 2000), 196–202; F. Ross Peterson, *Prophet without Honor: Glen H. Taylor and the Fight for American Liberalism* (Lexington, KY: University Press of Kentucky, 1974), 67–77, 161, 163.

nature-taming engineering projects in a way that may seem incongruous today, but that, in their own eyes, perfectly harmonized their longstanding ideals. Where, they might have asked, was the contradiction between an activist state managing the economy to ensure human welfare and a similarly activist one managing nature for the same purpose? No one, perhaps, had a better right to ask that question than Ernest Gruening (1887–1974). A Northeasterner by origin, he embodied the progressivism of the 1920s as managing editor of *The Nation*, publicity director for the La Follette campaign in 1924, and advocate for public hydroelectric power development. Interested, too, in foreign affairs, he was strongly opposed to war and imperialism, a friendly observer of anti-colonialist movements and the author in 1928 of a sympathetic study of Mexico in the aftermath of its revolution. He served the New Deal as head of the division of overseas territories and later as the appointed governor of Alaska. When statehood arrived in 1959, Gruening was chosen as one of Alaska's first US senators. He spent a decade as one of Congress's most liberal members and one of the first outspoken critics of the Vietnam War. Defeated for reelection in 1968, he was an early and enthusiastic supporter of George McGovern for the presidency in 1972.[63]

Gruening was also one of the Senate's most tireless critics of environmental and preservationist organizations that sought to block federal projects for developing the West's natural resources. The massive Rampart Dam, by which the Army Corps of Engineers proposed in the 1950s to create a Lake Erie-sized reservoir on the Yukon River in Alaska's interior, seemed to most impartial observers, in Marc Reisner's words, to make "no sense at all." Only Gruening's advocacy in the Senate kept the project alive for years.[64] In the autobiography he published in 1973, he couched his defense of the Rampart project, by then quite dead, in the familiar terms of early twentieth-century progressive Prometheanism. Its

[63] Ernest Gruening, *Many Battles: The Autobiography of Ernest Gruening* (New York, NY: Liveright, 1973); Robert David Johnson, *Ernest Gruening and the American Dissenting Tradition* (Cambridge, MA: Harvard University Press, 1998).

[64] Marc Reisner, *Cadillac Desert: The American West and Its Disappearing Water*, revised edition (New York, NY: Penguin, 1993), 210; Johnson, *Ernest Gruening*, 210–211, 214, 250–252; 272; Peter A. Coates, *The Trans-Alaska Pipeline Controversy: Technology, Conservation, and the Frontier* (Bethlehem, PA: Lehigh University Press, 1991), 134–138.

environmentalist opponents, he wrote, cared more for nature than for human beings—"Man requires a habitat too, and without a viable economy does he have one?"—and they failed to see that human intervention could greatly improve nature itself: "the Yukon Flats are, in my opinion, anything but beautiful. A great body of clear water over them could be an enhancement and supply a habitat for new varieties of wildlife."[65] As governor, Gruening argued for the eradication of wolves in McKinley National Park; as a senator, he strongly opposed federal plans to set aside large areas of Alaska as wildlife refuges. After visiting the USSR, he wrote enthusiastically about its massive programs of river control and development and about the ideals of social equality and human well-being that they symbolized.[66] In his autobiography, he praised Glen Canyon Dam on the Colorado River, of all western dams the one most abominated by preservationists, as an improvement on the natural scene that it had replaced: "where there was once a parched, uninhabited desert, there is now beautiful Lake Powell, with its clear and sparkling waters and facilities for fishing, swimming, and boating. … Who shall say that man's imagination and labor did not, in this instance, enhance unspoiled nature?" He defended the harvesting of timber in the Tongass National Forest in the tones of Gifford Pinchot: "scientific logging designed for perpetual yield is the essence of conservation."[67] (He did, however, fear the effects of population growth and fought as a senator for family planning.[68])

American Progressivism in the first half of the twentieth century, as well as being the name of three formal (though short-lived) political parties, also denoted a much wider movement of thought and action. To capture its spirit, one could do worse than take up the antithesis posed in the title of a book by a young enthusiast in 1914, *Drift and Mastery*, or the quest described by a famous synthesis of the era's history, *The Search*

---

[65] Gruening, *Many Battles*, 496–501 (quotations from 500).

[66] Roger Kaye, *Last Great Wilderness: The Campaign to Establish the Arctic National Wildlife Refuge* (Fairbanks, AK: University of Alaska Press, 2006), 26–31, 197–198; Coates, *The Trans-Alaska Pipeline Controversy*, 105–106, 138; Gruening, *Many Battles*, 414–415, 425–426, 430.

[67] Gruening, *Many Battles*, 500–501.

[68] Johnson, *Ernest Gruening*, 267–270.

for Order.[69] Progressives were all for mastery and order and all against drift. They had a high regard for science, engineering, expertise, rationality, control, planning, and efficiency, preferably deployed by a rational and benevolent government rather than a chaotic and cutthroat market, as necessary means to their favored ends, which included the reduction of economic inequalities and a systematic modernization of the ways of doing things that the past had bequeathed to the present. The Progressive spirit manifested itself in diverse forms in such areas as law, political science, philosophy, economics, education, public health, and historical scholarship, and when it touched upon the human relation to the earth, it was usually in the spirit of the progressive conservationists, seeking rational mastery over nature's forms and forces along with society's.

It suffused, for example, the work of the pioneering American sanitarian Ellen H. Richards (1842–1911). In 1904, Richards coined the word "euthenics" from Greek roots signifying "wellness" to denote "the science of controllable environment." She opposed euthenics to eugenics, or the applied science of raising human well-being by controlling human nature.[70] Eugenicists envisioned scientific experts speeding social progress by managing people's inherited makeup; euthenicists would pursue the same goal by correcting their surroundings. To Richards, the concept of surroundings or environment included not only the "natural" world, but the built and the social environment as well, but she saw society's very success to date in dealing with the first, "the power of nature-control which has been gained by mankind," as offering a model for emulation.[71] The clear improvements, to her mind, that humankind had effected in nature outdoors argued for the application of the same principles of efficiency and rationalization to the home: "man is civilized," she wrote, "in proportion as he dominates nature and bends hitherto unconquerable natural forces to minister to his needs." She looked to apply "those

[69] Walter Lippmann, *Drift and Mastery: An Attempt to Diagnose the Current Unrest* (New York, NY: Mitchell Kennerley, 1914); Robert H. Wiebe, *The Search for Order, 1877–1920* (New York, NY: Hill and Wang, 1967).

[70] Emma Seifert Weigley, "It Might Have Been Euthenics: The Lake Placid Conferences and the Home Economics Movement," *American Quarterly* 26, #1 (1974), 95–96.

[71] Ellen H. Richards, *Sanitation in Daily Life*, 2nd edition (Boston, MA: Whitcomb and Barrows, 1910), 59. She was quoting the English biologist E. Ray Lankester.

scientific principles which have spanned continents, controlled rivers, and tunnelled mountains to the building of houses that may be lived in safely and economically." Living spaces that were not kept up indeed had a natural tendency to decay, but:

> Is not one of the distinctive features of our age a forcible overcoming of the natural trend of things? If a river is by natural law wearing away its bank in a place we wish to keep, do we sit down and moan and say it is sad, but we cannot help it? No, that attitude belonged to the Middle Ages. We say, Hold fast, we cannot have that; and we cement the sides and confine or turn the water.[72]

Richards's program, in keeping with the Progressive outlook, called for intelligent government action of the sort that had already proven its value. "The state has applied science to ongoing problems, drained swamps, cleared out mosquitoes, furnished good water and drainage," and it could, she thought, equally well see to the quality of other key elements of the human environment.[73]

A coterie of American Progressive historians found the chief lesson of history to lie in the obsolescence of its legacies, of the institutions and ways of life created in less enlightened eras. No slavish admirers of the past that they studied, they emphasized its overall inferiority to the present, the progress that humankind had made over the course of time.[74] Their notion of progress included, and even centered on, the advances that had been made in the physical control of the earth and its processes. It certainly did for the foremost among them, Charles A. Beard (1874–1948), whose radical reputation was cemented by his 1913 book exploring the economic interests of the founding fathers in the US

[72] Ellen H. Richards, *The Cost of Living* (New York, NY: John Wiley & Sons, 1900), 21, 118, 156–157.

[73] Richards, *Sanitation in Daily Life*, 19; Ellen H. Richards, *The Art of Right Living* (Boston, MA: Whitcomb & Barrows, 1904), 35.

[74] For a classic expression, see James Harvey Robinson, *The New History: Essays Illustrating the Modern Historical Outlook* (New York, NY: Macmillan, 1912). On this group and its attitudes toward the past, see Peter Novick, *That Noble Dream: The "Objectivity Question" and the American Historical Profession* (New York, NY: Cambridge University Press, 1988), 92–100 and J.R. Pole, "The New History and the Sense of Social Purpose in American Historical Writing," *Transactions of the Royal Historical Society* 23 (1973), 221–242.

Constitution. Beard's earliest major work was a book-length essay titled *The Industrial Revolution* (1901), by which he meant both the episode in English history usually known by that name and the worldwide transformation of human life, potentially much for the better, effected by the machine. Through the development of technology, humankind had ceased to be "the helpless prey of the vulture elements" and had instead conquered and subdued them to its own service. Technology's potential to economize use and create substitutes allowed Beard to dismiss concerns about future energy shortages, and its ability "to fertilise and utilise the waste places of the earth; to render them more and more productive" likewise enabled him to make light of Malthusian fears regarding food supply.[75] In 1932, he pointed to the alleviation of the "many horrors wrought by physical nature" and "the subjugation of the material world to the requirements of human welfare" as the surest evidence that history was, on the whole, a success story. Treating "the substances and forms of nature as potential instruments of human progress," he thought, would be the surest way to make it continue. Writing at the depth of the Great Depression, he emphasized more strongly than ever that progress was "necessarily planful," that it could not be expected to occur on its own without conscious direction.[76]

The wide-ranging work of the philosopher John Dewey (1859–1952), both as a thinker and as a political activist, embodied many of the key themes of Progressivism. It would be a huge misunderstanding of Dewey, as indeed of Beard, to describe him as a technocrat in the usual sense of the word, as one who sees technical expertise best deployed when free of the annoyances and interferences of democratic politics. His philosophy, all the same, is fundamentally one of technology, of the intelligent enlargement of the means by which people may satisfy their wants (and at the same time of their conception and selection of wants themselves).[77]

---

[75] Charles A. Beard, *The Industrial Revolution* (London: Swann, Sonnenschein & Co., 1901), 1, 93, 98.

[76] Charles A. Beard, "The Idea of Progress," in Charles A. Beard, ed., *A Century of Progress* (New York, NY: Harper & Brothers, 1932), 4, 6, 18.

[77] For a discussion, see Larry A. Hickman, *Dewey's Pragmatic Technology* (Bloomington, IN: Indiana University Press, 1990).

Finding the term "pragmatism" too subject to misunderstanding, Dewey preferred "instrumentalism" as a label for his approach. Thought, for him, was not a means of discerning and contemplating eternal verities or the essences of things; it was an instrument like any other, though more powerful than most, for rendering life less hazardous and more satisfying by controlling and improving its environment. Logic itself was not communion with timeless and transcendent truths, but an empirical and progressive science of the tools of successful problem-solving.[78] Dewey saw the problem-solving achievements of the engineer as a model for bringing the forces of social life under control, deploring the kind of conservatism that he defined as "disbelief in the possibility of constructive social engineering."[79] To the possibilities of environmental engineering he was no less receptive. A number of writers in recent years have interpreted Dewey's works as potentially important contributions to the philosophy of environmentalism.[80] Yet, their environmental imagery is largely Promethean in character, just as the overt theme of his entire philosophy is the control of human beings' surroundings by their intellect to better serve their wants.

Critical of most earlier philosophers for leading their successors down a number of blind alleys, Dewey lauded Francis Bacon as the first to have pointed towards a better approach. Bacon, he wrote in 1920, was "the great forerunner of the spirit of modern life" and indeed "the real father of modern thought." He had been the first to break fundamentally with the erroneous traditions of Western philosophy, with whose self-absorbed futility Dewey contrasted the fruitfulness of Bacon's approach,

---

[78] John Dewey, *The Quest for Certainty* (orig. 1929), in Jo Ann Boydston, ed., *The Later Works, 1925–1953*, vol. 4 (Carbondale and Edwardsville, IL: Southern Illinois University Press, 1984); *Logic: The Theory of Inquiry* (orig. 1938), in ibid., vol. 12, 1986.

[79] John Dewey, "Progress" (orig. 1916), in Jo Ann Boydston, ed., *The Middle Works, 1899–1924*, vol. 10 (1980), 241.

[80] William Chaloupka, "John Dewey's Social Aesthetics as a Precedent for Environmental Thought," *Environmental Ethics* 9, #3 (1987), 243–260; Hugh P. McDonald, *John Dewey and Environmental Philosophy* (Albany, NY: SUNY Press, 2004); Ben A. Minteer, *Reformulating Environmental Ethics: Pragmatism, Principle, and Practice* (Philadelphia, PA: Temple University Press, 2012); though cf. Bob Pepperman Taylor, "John Dewey and Environmental Thought," *Environmental Ethics* 12, #2 (1990), 175–184. Of course it does not follow that his philosophy is necessarily or inherently Promethean in environmental matters, merely that he himself applied it in that way in his own time.

"the clearness and the force with which he associated both the need and the possibility of progress, to be brought about through a scientific knowledge of natural conditions and taking effect in inventions directed toward ameliorating the lot of man." The contemplative philosophy of the past had led nowhere. The applied natural science heralded by Bacon had led to "the Empire, as he says, of Man over Nature, substituted for the Empire of Man over Man."[81] Technology, or applied thought, Dewey wrote, was the means by which human beings were enabled to "attack and transform nature." He echoed Saint-Simon, knowingly or not: "The Golden Age lies ahead of us not behind us." He repeatedly referred to the human control or command or conquest or mastery of nature as a key measure of civilization, indeed as the surest basis by far for a belief in progress.[82] In the same vein, Dewey's fellow pragmatist and political progressive William James in 1910 published an appeal for conscription into a new kind of army to wage "the moral equivalent of war," which he defined as "the immemorial human warfare against nature," "the army enlisted against *Nature*," its task to make "this only partly hospitable globe" a proper home for humankind.[83]

Dewey's major works, even when dealing with other matters, conveyed the same Promethean message in their use of transformations of the environment as symbols of human advances. The savage, Dewey wrote, lives in a wilderness on the productions of nature and as a result "leads a meagre and precarious existence." On the other hand:

> The civilized man goes to distant mountains and dams streams. He builds reservoirs, digs channels, and conducts water to what had been a desert. He searches the world to find plants and animals that will thrive. He takes native plants and by selection and cross-fertilization improves them. He

[81] John Dewey, *Reconstruction in Philosophy* (orig. 1920), in *The Middle Works*, vol. 12, 95, 100; "Contributions to *Cyclopedia of Education*" (orig. 1912–1913), in ibid., vol. 7, 332.

[82] Dewey, *Reconstruction in Philosophy*, 106, 120; *Democracy and Education* (orig. 1916), in *The Middle Works*, vol. 9 (1980), 219, 223, 224.

[83] William James, "The Moral Equivalent of War" (orig. 1910), in *William James: Writings, 1902–1910*, ed. Bruce Kuklick (New York, NY: Library of America, 1999), 1291.

introduces machinery to till the soil and care for the harvest. By such means he may succeed in making the desert blossom like the rose.[84]

Elsewhere and just as approvingly, Dewey likened the advance of civilization to the replacement of rough wilderness paths with smoothed roads. Expecting changes in individuals' hearts and minds to solve social problems was like "supposing that flowers can be raised in a desert or motor cars run in a jungle. Both things can happen and without a miracle. But only by first changing the jungle and desert." Individual wishes, he granted, mattered, but only when they led to more effective action: "Taste for flowers may be the initial step in building reservoirs and irrigation canals," or "the work that makes the desert blossom." He clothed a call for expanded state action in a Promethean metaphor: agencies were needed "which canalize the streams of social action and thereby regulate them." Dewey rejected the reverence for nature he saw in many other philosophers. He equated it with "supposing that knowledge of the connection between malaria and mosquitoes enjoins breeding mosquitoes," instead of using the knowledge to disrupt the natural pattern "through draining and oiling swamps, etc." and to replace it with something better.[85]

Finally, Dewey had, too, the characteristic progressive and Promethean skepticism of limits and scarcity as inexorable facts of nature. He saw them, rather, as questionable alibis for the persistence of poverty in a world of potential abundance. He defined natural resources not as materials good by their nature, but as things created when intelligent action exploited the potentialities within neutral matter to fashion means for attaining human ends. He noted that society still made use of only a small, though growing, range of nature's possibilities in this way. He pointed to the replacement of organic manure by chemical fertilizers as illustrating the promise of "substitution and convertability" to satisfy human

---

[84] Dewey, *Reconstruction in Philosophy*, 128–129; for a similar statement, see *Democracy and Education*, 52.

[85] John Dewey, *Human Nature and Conduct* (orig. 1922), in *The Middle Works*, vol. 12 (1982), 20, 206; *The Public and Its Problems* (orig. 1927), in *The Later Works*, vol. 2 (1984), 317, 358 (see also 268, 269–270, 339).

demands on the natural world.[86] Such arguments formed the core of American Progressive resource economics. If the Malthusian conception of finite natural resources as a limiting factor for society is a basic theme of environmentalism, that of natural resources as human creations, and therefore not necessarily finite, is a characteristic one of Prometheanism. Scarcity, this view implies, is a social creation, and an alterable one, not an inexorable consequence of nature's limits. Adumbrated by some of the progressive conservationists in the first decade of the twentieth century, it was developed in full in the next several decades by several heterodox economists, the pioneers of American institutionalism. Each associated the doctrine with a skeptical attitude toward unregulated market capitalism and toward the orthodox economic theory that took capitalism's merits for granted.

In his 1921 book *The Engineers and the Price System*, Thorstein Veblen (1857–1929) favorably contrasted the technocratic principle, a devotion to rationality, efficiency and planning, with the inefficiencies of the market or pecuniary economy prevalent in America.[87] Two years later, he drew attention to the role of technology in creating natural resources. The USA, he observed, was a land notably abundant in such resources. But their abundance, he continued, was not a fact of nature but rather the product of human actions. The "industrial arts," as Veblen called them, had "converted otherwise meaningless elements of physiography and mineralogy into industrial wealth." The USA was unusually rich in what counted by world standards as resources, because its institutions had given the industrial arts an unusual freedom to develop, one that, understandably enough, they had devoted to finding uses for the substances that the country possessed in the most plenty. Just as technology had made resources, Veblen argued, further developments in it could make them, though physically unaltered, cease to be resources. Resource scarcity and abundance, then, were not given, unalterable natural conditions, but the work of society.[88] Veblen stood almost alone among the

[86] Dewey, *Reconstruction in Philosophy*, 120; "Nature and Its Good: A Conversation" (orig. 1909), in *The Middle Works*, vol. 4 (1977), 16, 20.

[87] Thorstein Veblen, *The Engineers and the Price System* (New York, NY: B. W. Huebsch, 1921).

[88] Thorstein Veblen, *Absentee Ownership and Business Enterprise in Recent Times* (New York, NY: B. W. Huebsch, 1923), 124, 166–167, 272.

leading American economists of his era in his skeptical attitude towards the Malthusian theory of population. He regarded the natural limits that it postulated as, in fact, socially produced, the result of entrusting production to an inefficient scheme of laissez-faire rather than to one of technocratic planning that could turn scarcity into abundance.[89]

What Veblen sketched out, a concept of natural resources as human creations, the German-born, America-based economist Erich W. Zimmermann (1888–1961) developed in detail. As opposed to a prevalent "physical theory" of natural resources, he expounded a "functional theory" of resources as generated not by nature but by human intellect through the development of technology. After applying the theory to the world's major resource sectors, Zimmermann discussed its relation to problems of political-economic organization and emphasized the importance of direction by the state. Private owners, he argued, had a vested interest in keeping resources scarce and therefore costly, the public in making them abundant and cheap; hence the processes of resource development should be vested in the public sector. This called, among other things, for "world-wide economic planning" rather than "a passive reliance on the mysterious workings of so-called economic laws."[90] The American institutionalist economist Clarence W. Ayres (1891–1972) put forward much the same interpretation of natural resources, a term he criticized as a misnomer. They were not, he maintained, natural phenomena, nor was scarcity a fact of nature. Technology created resources. The erroneous assumption of their natural and finite character was a prop supporting the worldview of market capitalism, one Ayres regarded with deep skepticism.[91] So did his associate Walton Hamilton (1881–1958), who first applied the label "institutionalist" to their approach. In 1944, Hamilton stated in a few sentences the same understanding of the created character and the indefinite and perhaps infinite quantity of natural resources that became the mainstay in later decades of Julian

[89] Joseph J. Spengler, "Veblen on Population and Resources," *Social Science Quarterly* 52, #4 (1972), 861–78.

[90] Erich W. Zimmermann, *World Resources and Industries: A Functional Appraisal of the Availability of Agricultural and Industrial Resources* (New York, NY: Harper & Brothers, 1933), 801, 807.

[91] Clarence E. Ayres, *The Theory of Economic Progress* (Chapel Hill, NC: University of North Carolina Press, 1944), 84, 113.

Simon's market-based Prometheanism: "New sources of supply may be discovered; new products may appear as substitute materials; materials equivalent in office and character may emerge from a novel process. It is technology which gives value to the stuffs which it processes; and as the useful arts advance the gifts of nature are remade."[92]

The American economist Simon N. Patten (1852–1922), a pioneer of some of the themes of the institutionalists, had earlier proclaimed a new era of technology-created plenty. "Do resources diminish with use?," he asked in 1907, and his answer was no; "a stupendous change" was being brought about by the advance of science and industry, though hindered by the persistence of customs and institutions inherited from the age of scarcity, including a competitive market economy. "Machinery, science, and intelligence moving on the face of the earth" had superseded natural forces as its chief shaping powers, Patten argued, and were making the land ever more productive and ever more hospitable to human occupance.[93] The Great Depression, in the eyes of many, made the contrast between the seeming potential of abundance and the reality of want particularly intolerable. Technocracy became a proper noun in the early 1930s as the adopted name of a Depression-inspired American political movement emphasizing the purely institutional character of scarcity in a world where technology had overcome the natural limits of resources, particularly of energy, and needed only the expert and disinterested guidance of engineers to make a good life available to all.[94]

In its concern with efficiency and with the direction of the trained engineer as the key to achieving it, American technocracy—the conservation movement included—had a close affinity to the work of the pioneering production engineer Frederick Winslow Taylor.[95] Taylor's influence had a worldwide reach, and one of those who felt it was the Russian

[92] Walton Hamilton, "The Control of Strategic Materials," *American Economic Review* 34, #2, pt. 1 (1944), 262.

[93] Simon N. Patten, *The New Basis of Civilization* (New York, NY: Macmillan, 1907), 13, 25–26. On Patten's life and thought, see Daniel M. Fox, *The Discovery of Abundance: Simon N. Patten and the Transformation of Social Theory* (Ithaca, NY: Cornell University Press, 1967).

[94] William Ernest Akin, *Technocracy and the American Dream: The Technocrat Movement, 1900–1941* (Berkeley, CA: University of California Press, 1977).

[95] Steinberg, *Down to Earth*, 139, notes the parallel between Taylorism and the conservation movement.

Aleksei K. Gastev (1882–1939). Enamored of industrial technology as the essence of modernity, Gastev blended Taylorism with the political radicalism that he espoused as a revolutionary student and as a member of the radical Bolshevik wing of Russian Marxism. Exiled or self-exiled a number of times before 1917, he subsequently devoted himself to the cause of reconciling human labor and mechanization, one he pursued in visionary ways as head of the new Soviet state's Central Labor Institute before falling afoul of Stalin's purges in the next decade.[96] The literary work that he largely abandoned after 1917 expressed perhaps the most exultant technological Prometheanism of any writer of note. Gastev was the prose poet of the artificial landscape and the industrial scene, of the obliteration of the natural environment by a human race that has seized nature's powers and wields them to its own purposes. In Gastev's world—for it is one world, its natural barriers overcome and all of the earth interconnected by rail and tunnels and bridges—forests are cleared, marshes drained, and rivers confined. They are replaced by industrial cities, floors of concrete crisscrossed by steel rails, dominated by vast factories and powerful machines, with cranes and towers reaching high above and the explosions of blasting reaching deep down into the earth below, mingled with the triumphant racket of hammer, whistles, and trains. The forces of socialist construction are remaking the very planet, and to enhance their work Gastev dreamed of transforming the laboring human body itself into a machine.[97]

A still more prominent technocratic Promethean of the Russian revolutionary movement was A.A. Bogdanov (born A.A. Malinovskii, 1873–1928). A physician by training, Bogdanov as a young man joined the

[96] Kendall E. Bailes, "Aleksei Gastev and the Soviet Controversy over Taylorism," *Soviet Studies* 29, #3 (1977), 373–394; Richard Stites, *Revolutionary Dreams: Utopian Vision and Experimental Life in the Russian Revolution* (New York, NY: Oxford University Press, 1989), Chap. 7; Rolf Hellebust, *Flesh to Metal: Soviet Literature and the Alchemy of Revolution* (Ithaca, NY: Cornell University Press, 2003). Though Russian Bolshevik thought had had a strong technocratic strain, under Stalin engineers were put firmly in their place: Kendall E. Bailes, "The Politics of Technology: Stalin and Technocratic Thinking Among Soviet Engineers," *American Historical Review* 79, #2 (1974), 445–469.

[97] Aleksei Gastev, *Poeziia rabochego udara* (Moscow: Sovetskii pisatel', 1964), especially the prose poems "Bashnia" (102–106), "Rel'sy" (107–108), "Kran" (109–112), "My posiagnuli" (129–131), "My vmeste" (132–134), "Express" (135–152), "Vorota zemli" (164–165), "Chudesa raboty" (170–175), and "Vykhodi" (187–188).

ranks of the Bolsheviks, only to be expelled after losing an ideological and power struggle with his adversary Lenin. Living mostly in political exile after the failed revolution of 1905, he returned to Russia in 1913 and, following the Revolution, taught and wrote in Moscow until his death. Both before and after 1917, Bogdanov devoted much effort to developing a system of thought and social organization that he dubbed "tectology," which emphasized the importance and coordination within a society of expert technical knowledge. Critical of the undemocratic and hierarchical traits of the Soviet regime, he argued for greater decentralization of power and particularly for more autonomy for engineers, experts, and workers in their own spheres of production. He was the founder and leading spirit shortly after the Revolution of the Proletkult movement, which sought to formulate new modes of culture, science, education, and everyday life appropriate to new and revolutionary times in place of those inherited from the past.[98]

In harmony with orthodox Marxism, Bogdanov in his nonfiction writings emphasized the struggle against and progressive mastery of nature as humankind's most important task. Much like Dewey, he treated knowledge and inquiry as instruments to human ends, and like Dewey he emphasized the roots of knowledge in experience and the necessarily social character of reliable knowledge. Fusing a worldview and an earthview, he disparaged "elemental conservatism," assimilating the forces of tradition, capitalism, and the status quo to the mindless forces of nature and those of radical socialist revolution to their mastery by rational planning. The essential character of post-capitalist society, he asserted, would lie in the ever-developing power of society over nature achieved through organized applied science. As one of the principal achievements and further tools of such power, he envisioned sources of energy (not least from

[98] The fullest English account of Bogdanov's life, though selective in its focus, is Nikolai Krementsov, *A Martian Stranded on Earth: Alexander Bogdanov, Blood Transfusions, and Proletarian Science* (Chicago, IL: University of Chicago Press, 2011). On his prewar thought, see also Zenovia A. Sochor, *Revolution and Culture: The Bogdanov-Lenin Controversy* (Ithaca, NY: Cornell University Press, 1988).

the atom) dwarfing in their magnitude all others that human society had previously possessed.[99]

Bogdanov dramatized this ideal, and the cult of the socially engaged engineer that was for him its logical corollary, in a novel, *Krasnaia zvezda* (*Red Star*), published in 1908, and in a prequel, *Inzhener Menni* (*Engineer Menni*), which appeared five years later. The two novels—both of them popular and widely read in Russia before and after 1917—recount what a visitor from Earth learned by observation and conversation regarding the planet Mars and its inhabitants.[100] For their scientific background, Bogdanov drew on some theories then much in the news. Near the end of the nineteenth century, the American astronomer Percival Lowell claimed to be able to see, through the telescope of his observatory at Flagstaff, Arizona, long lines running across the Martian surface. Continuing his researches for the next two decades, Lowell mapped a network of such lines and interpreted them as most representing likely bands of vegetation running along irrigation canals that the intelligent inhabitants of a slowly drying planet had built to channel water from the seasonal melting of their polar ice caps.[101] Both of Bogdanov's novels built on this foundation to suggest what political organization and historical development must have underpinned such world-scale management of the environment—and by implication, what must be done on Earth if similar Promethean feats were ever to be possible there.

The central episode in the history of Mars as both books recount it was that of "the Great Project," whose unfolding occupies most of *Engineer Menni* and helped give rise to the society depicted in *Red Star*. After centuries of social development and the steady consolidation of

---

[99] A.A. Bogdanov, *Novyi mir* [orig. 1904; in *Voprosy sotsializma: raboty raznykh let* (Moscow: Izdatel'stvo politicheskoi literatury, 1990)], 60–61 (see also 48–49, 51, 58–59); "Sotsialisticheskoe obshchestvo" (orig. 1906; in ibid.), 90–98; "Voprosy sotsializma" (orig. 1918; in ibid.), 297–298.

[100] Alexander Bogdanov, *Red Star, The First Bolshevik Utopia: Red Star, Engineer Menni, A Martian Stranded on Earth*, trans. Charles Rougle, ed. Loren R. Graham and Richard Stites (Bloomington, IN: Indiana University Press, 1984). On the book's background and reception, see Richard Stites, "Fantasy and Revolution: Alexander Bogdanov and the Origins of Bolshevik Science Fiction," in ibid., 1–16.

[101] William Graves Hoyt, *Lowell and Mars* (Tucson, AZ: University of Arizona Press, 1976); Michael J. Crowe, *The Extraterrestrial Life Debate, 1750–1900: The Idea of a Plurality of Worlds from Kant to Lowell* (Cambridge: Cambridge University Press, 1986).

once-separate countries, war had broken out between reactionary and progressive forces. The last great feudal warlord of Mars had been finally defeated and killed and a republic spanning the planet's surface firmly established. But that surface itself remained to be unconquered, more than half of it uninhabitable arid wasteland or pestilential swamps and bogs. Enter the young engineer Menni, a committed republican though the son of the last great feudal leader. He drew up a scheme to blast a channel through a mountain chain and open a route for the ocean to flow into a vast desert and create an inland sea. Its waters would so modify the once-arid climate that the rich soils would become cultivable with the help of supplemental irrigation and far more than suffice to repay the cost of the work. The scheme was approved and Menni carried it out according to plan. "The elements had been dealt a tremendous blow, and it began to seem as though man could accomplish anything he set his will to." Menni then devised and was able to win a hearing for his larger project, "which foresaw the transformation of the entire planet," the conquest and irrigated cultivation of all of its wastelands.[102] A work of environmental improvement and of economic stabilization at once, it would prevent an impending depression by stimulating the economy and assuring future prosperity through the income that the reclaimed land would generate. But mere technical knowledge came into collision with the new economic realities of capitalist inequality: the discontent of the labor force over their poor working conditions, and the near-disappearance of independent smallholding farmers as land ownership became more and more concentrated in the hands of a few. In the crisis that resulted, Menni's son, who grasped as his father did not the links between social and environmental transformation, showed the way forward.

In the socialist utopia that resulted, the one depicted in *Red Star*, the canal system irrigates the land, and industry is efficiently organized for production and for the mastery of nature. As population and the demand on resources grow, though, periodically they threaten the natural resource base: "the tighter our humanity closes ranks to conquer nature, the tighter the elements close theirs to avenge the victory." But the Martians do not respond by conceding defeat and controlling their numbers and

[102] Bogdanov, *Red Star*, 164–165.

curbing their growth; they resort in the classically Promethean way to science and technology for the answer and have always done so successfully. In the past, they have dispelled crises of energy and of metal ore supplies by developing substitute sources. At present, food supply is the limiting resource, and expansion to other worlds a possible solution. The Martians reject the colonization of Earth, on the grounds of the injustice it would do to that planet's inhabitants, and they project that of Venus instead. "We can triumph as long as we are on the offensive," one tells a visitor from Earth, "but if we do not permit our army to grow, we will be besieged on all sides by the elements, and that will in turn weaken faith in our collective strength, in our great common life."[103] Surrendering to nature's limits, rather than perpetually challenging them, would mean the surrender of what their society stands for, and what their creator believed in, too.

---

[103] Ibid., 79–80 (quotations), 109–121.

3

The Scientific Prometheans: Studying Nature to Improve It

Bogdanov was not alone in his time in looking to Mars for hints of what humankind might one day make of its own environment. The American Lester Frank Ward (1841–1913), for most of his career a biologist by profession and a sociologist by avocation, found Percival Lowell's picture of the red planet's Promethean transformation as attractive a topic for speculation and inference as the Russian revolutionary had. This neighboring world seemed to harbor a race of beings capable of effecting transformations large enough to be seen across the distances of space. Their apparent achievement tempted Ward, like Bogdanov, to wonder what kind of society could have accomplished such feats and what lessons could be drawn for the earth.

Even before Lowell, the Italian astronomer Schiaparelli had reported the existence of long straight lines running across the Martian surface and had speculated that they might be the work of intelligent creatures.[1] In 1893, Ward had seized on the idea to exemplify a favorite thesis of his: the vast superiority of rational purpose over the haphazardness of mindless

[1] Crowe, *The Extraterrestrial Life Debate, 1750–1900*, 500–502, 514–515.

W.B. Meyer, *The Progressive Environmental Prometheans*,
DOI 10.1007/978-3-319-29263-2_3

and uncontrolled nature. Such lines as Schiaparelli had described, he wrote, would indeed be signs of conscious intelligence. "Nature has no economy. Only through foresight and design can anything be done economically. Rivers thus constructed (canals, mill-races, irrigating ditches, etc.) are straight, or as nearly so as true economy requires"; one could always know (and despise) a natural stream by the way it meandered shiftlessly toward its outlet.[2]

His research on fossil plants brought Ward in June of 1901 to Flagstaff, Arizona, where Lowell's observatory was located. One evening he accepted an invitation to look at Mars through the main telescope and record his impressions. He saw what his host and others had seen, "great canals or long cavities in various directions," the supposed bands of vegetation bordering the lines of the irrigation network across the planet.[3] Having seen, he believed. Satisfied that rational beings had indeed reshaped the Martian landscape, Ward several times again held up their supposed achievement to his fellow earthlings as a Promethean paradigm: a case of intelligent creatures who had united to engineer their environment on a stupendous scale. In 1903, he contrasted the size of the Martians' achievement with the difficulties, then still unconquered, that humankind had experienced in so comparatively small a project as the opening of a Panama Canal. "Man has only just begun the conquest of nature," Ward wrote. "We may suppose that in Mars the conquest of nature is complete and that every law and every force of nature has been discovered and utilized. Under such conditions there would seem to be scarcely any limit to the power of the being possessing this knowledge to transform the planet and adapt it to its needs." Nor, he made it clear, did he think there ought to be any such limit. On earth, too, man would eventually "become absolute master of his physical environment," such that "the operations which he now performs will seem like the work of ants."[4] In 1907, after reading Lowell's first two books on the planet, Ward published an article on "Mars and Its Lesson": one that, by his own account, received more attention than any

[2] Lester F. Ward, *The Psychic Factors of Civilization* (Boston, MA: Ginn & Company, 1893), 256.

[3] Lester F. Ward, *Glimpses of the Cosmos*, vol. 6 (New York, NY: G. P. Putnam's Sons, 1918), 244.

[4] Lester F. Ward, "Social Differentiation and Social Integration," *American Journal of Sociology* 8, #6 (1903), 744–745; *Pure Sociology: A Treatise on the Origin and Spontaneous Development of Society* (New York, NY: Macmillan, 1903), 470–471.

other that he had written. Mars's lesson, as Ward read it, lay in what it said about the earth. It showed that in the normal course of its development, a planet must give rise to intelligent life, and intelligence would in due course come to direct its further development: "man has all but ceased to feel the transforming influence of his environment, and … has begun, on a grand scale, to transform his environment." In conclusion, Ward approvingly quoted Saint-Simon: the golden age for humankind lies ahead, and not in the past.[5]

It is not surprising that engineers, or technocrats enamored of engineering, should have possessed a frankly Promethean outlook on environmental matters. What is engineering, what is technology, if not the alteration of nature? From practitioners of the pure, as opposed to the applied and technical, sciences, though, one might well expect something different. Their task, one could suppose, is to study the order of nature and not to remake it. If the work gives them any particular bias, it might seem to be one towards preservation. In the Western world today, natural scientists do appear as a rule to be more than ordinarily skeptical of the wisdom and benefits of Promethean environmental alteration. But the rule may not have held true in the nineteenth and early twentieth centuries. Some elements common to the experience of many scientists of the time came together for them into a coherent rationale for the transformation of nature rather than its maintenance.

If Promethean technocrats and engineers emphasized the tools of human invention that made the reshaping of physical geography possible, their scientific counterparts focused for their part on the defects in nature that made such reshaping desirable. A close acquaintance with its operations made many of them aware of nature's prodigal purposelessness and wastefulness, compared with the efficiency that rational human beings could impose. It was an insight they shared with engineers, but it extended to realms, such as plant and animal life and the workings of the atmosphere and the cosmos, with which engineers of their day rarely dealt. Especially after 1859, too, scientists persuaded by Darwin's account of evolution were much more disposed than their predecessors

[5] Ward, *Glimpses of the Cosmos*, vol. 6, 244–245, 247; "Mars and Its Lesson," *The Brown Alumni Monthly* 7, #8 (1907), 159–165.

had been to note, not the wonders of design in the features of the earth and its forms of life, but the evidence of a lack of design in its multitude of imperfect adaptations. The historical sciences of biology and geology combined as well to underline the transience of nature's configurations and to take away some of the sacredness and untouchability that their supposed divine origin or their seeming permanence in the present day had given them. Both disciplines also made apparent the poverty of nature under its apparent richness, the many species and features that could exist but did not; on close inspection, the "great chain of being" turned out to consist mostly of gaps. Among the absentees were many species that had left fossil traces but were no longer to be found alive. Some scientists could not help drawing an ominous lesson: that the human race might prove just as transient as any of these extinct creatures if it relied on nature's caprices as the sole guarantee of its survival. The evidence of the longer term only reinforced what the workings of disease and geophysical calamity suggested every day: nature's entire indifference to human persistence and well-being. And if the pragmatic concerns of self-defense pointed away from a meek submission to nature and its laws, the loftier ones of morality did the same. Nature, examined closely, tolerated and, indeed, seemed to reward an appalling degree of cruelty and unfairness among living creatures. To Charles Darwin, who understood it better than anyone else in his century, it had not even the excuse of efficiency for the outrages that it perpetrated; he summed up its workings as "clumsy, wasteful, blundering low & horridly cruel."[6] Darwin's devoted disciple T.H. Huxley urged humankind not to look to nature for lessons in conduct, lest they fall to mimicking its unconscionable ways: "the ethical progress of society," he wrote, did not involve imitating nature but rather "combating it."[7] By the application of intelligence and moral insight, human beings had already "succeeded in building up an artificial world within the cosmos" and achieved "a command over the course of non-human nature greater than that once attributed to the magicians."[7]

[6] Charles Darwin to J.D. Hooker, 13 July 1856, in Frederick Burkhardt and Sydney Smith, eds., *The Correspondence of Charles Darwin*, vol. 6: 1857–1857 (Cambridge: Cambridge University Press, 1990), 178.

[7] Thomas H. Huxley, *Evolution and Ethics and Other Essays* (New York, NY: D. Appleton and Company, 1897), 83, 84.

Just as the American progressive historians on the whole disliked the past that they studied, some natural scientists concluded that human beings not only ought not to follow nature, but should try to change it.

These turns of thought paralleled others that their training inclined some scientists to make about the world of human society. The practice of research sharpened a habit of criticism that disposed them to be skeptical of the value of mere tradition and of the existing arrangements in political, social, and economic life that it was used to justify. For this reason, the rabid anti-traditionalists of the eighteenth-century Enlightenment had found science particularly attractive both as a pursuit and as a symbol. It had seemed to them inherently radical, demystifying custom and practice and the supposed wisdom embodied in tradition that Burke, more than anything else, made the essence of conservative philosophy. And at least in some circles, science in the nineteenth century acquired an additional aura of radicalism from the doubts that geology and biology, through their reconstruction of the past, and chemistry and physics, through their advances in purely material explanations of phenomena, cast upon the traditional doctrines of a revealed Christian religion that had entered into a partnership with other agencies of social order and stability. The idea of a warfare between science and religion, or between science and theology, at once applied to the present and projected onto the past, acquired a wide currency that it had not before possessed.[8] The habit of questioning often extended itself to such other established components of the social order as the institutions of government, the roles of the sexes, and the mechanisms of economic production and consumption. Those scientists who challenged traditional teachings in these areas often suggested, more or less openly, that the arrangements that had sprung up in the less knowledgeable past and been handed down to the present could benefit from a good deal of intelligent redesign or even be replaced wholesale by ones more efficient and more appropriate to changed times.

---

[8] John William Draper, *History of the Conflict between Religion and Science* (New York, NY: D. Appleton and Company, 1874); Andrew D. White, *A History of the Warfare of Science with Theology in Christendom* (New York, NY: D. Appleton and Company, 1896). On their role in propagating the idea, see John Hedley Brooke, *Science and Religion: Some Historical Perspectives* (Cambridge: Cambridge University Press, 1991), 33–42 and Peter Harrison, *The Territories of Science and Religion* (Chicago, IL: University of Chicago Press, 2015), 171–175.

Ward represents a first-rate specimen for study of the species in question, the progressive scientific Promethean. An Illinois farm boy, a veteran of the American Civil War, largely self-taught as a scientist, he spent most of his working life as a paleobotanist at the Smithsonian Institution and the US Geological Survey in Washington. On his own time and with the connivance of sympathetic supervisors, he made a second career of speculative social thought, ending his life as the occupant of a new chair in sociology at Brown University.[9] His first important work of social science, *Dynamic Sociology*, appeared in two volumes in 1883. He followed it in 1884 with a short article, "Mind as a Social Factor," and a bristlingly hostile review of a new book by William Graham Sumner, the Yale sociologist and chief American advocate of what has come to be known as Social Darwinism.[10] These works set out all of the themes that would be central to Ward's voluminous later writings.

Chief among them was a contrast between what Ward called the natural and the artificial, between the workings of non-human nature and those of rational human action, a contrast very much to the former's disadvantage. Mingling geology and biology, his research on fossil plants made Ward aware as few others were of the vast stretches of time that the natural evolution of species had required. Nature worked slowly, blindly, and with an appalling wastefulness, one best seen in the huge redundancies involved in plant and animal reproduction in the wild. A spawning female herring, Ward noted, released ten thousand eggs; a "large chestnut tree in July," a ton of pollen grains.

> Nature's way of sowing seed is to leave it to the wind, the water, the birds and animals. The greater part falls in a mass close to the parent plant and is shaded out or choked to death by its own abundance. Only the few seeds that chance to be transported by one agency or another to some favorable spot and further happen to be covered up, can grow. The most of them that

[9] By far the best account of Ward's life and thought is Edward C. Rafferty, *Apostle of Human Progress: Lester Frank Ward and American Political Thought, 1841–1913* (Lanham, MD: Rowman & Littlefield, 2003).

[10] Lester F. Ward, *Dynamic Sociology, or Applied Social Science*, 2 vols. (New York, NY: D. Appleton, 1883); "Mind as a Social Factor," *Mind* 9, #36 (1884), 563–573; "Professor Sumner's Social Classes" (orig. 1884), reprinted in *Glimpses of the Cosmos*, vol. 3 (New York, NY: G. P. Putnam's Sons, 1913), 301–305.

> germinate never attain maturity on account of hostile surroundings, and only the rarest accidents of fortune live long enough to continue the race. To meet this enormous waste correspondingly enormous quantities of seed are produced. How different the economy of a rational being![11]

Against the tradition of emphasizing the wisdom displayed in the intricate designs of nature, Ward drew attention to the many ways in which it was badly made and in which its present patterns were ill-designed to meet human wants or even the maximum flourishing of other species.[12] The perfect adaptation of living things to their surroundings, he asserted, a frequent motif both of the old religious argument from design and of the emerging worldview of Social Darwinism, was an illusion. Many plants flourished much more abundantly when relocated from their native settings to new habitats. "The adaptations of Nature of which we hear so much are *not* perfect."[13] They had arisen, with all of the enormous inefficiency that they entailed, because of the absence of a rational mind that could match the right means to its chosen ends.

The earth's surface, Ward observed, had demonstrably not been designed in the best interests of the human race. Its history, too, showed that the forces that had shaped it knew and cared no more about the well-being or even survival of people than about those of any other species. And if nature's workmanship left much to be desired, Ward added, even more so did its morals, which were abhorrent and unsuitable for human emulation.[14] The rest of sentient nature, he argued, the animal kingdom in the aggregate, was made better off by the human interference with a natural order that was one of "chronic terror" for "every living thing." Human beings protected domesticated animals, thinning or exterminating their wild and carnivorous enemies. An inventory of their net effect suggested to Ward "that a larger amount of animal life exists under man's influence than could exist without it; that he creates more life than he destroys; that his methods of destruction are less painful than those of

[11] Ward, *The Psychic Factors of Civilization*, 247, 256.

[12] Ward, *Dynamic Sociology*, vol. 2, 64–71.

[13] Lester F. Ward, "The Local Distribution of Plants and the Theory of Adaptation" (orig. 1875), reprinted in *Glimpses of the Cosmos*, vol. 2 (New York, NY: G. P. Putnam's Sons, 1913), 39–40, 43.

[14] E.g., Ward, *Dynamic Sociology*, vol. 1, 503–504, 524; "Mind as a Social Factor," 570.

Nature; that it is to his interest to treat animals well, to supply them with abundant food, and relieve them from those constant fears, both of enemies and of want, which characterize their condition in a wild state; and that when life is taken, it is done quickly and as painlessly as possible; that the reverse of all this is the case in Nature."[15]

Looking at things in this light, Ward marveled at the prevalence of "nature-worship" ("physiolatry" he would also call it, in one of the neologisms he was fond of coining) among the leading philosophers and social thinkers of his day, particularly the Englishman Herbert Spencer and the American Sumner and their disciples, who "devote their energies to extolling the natural and decrying the artificial."[16] Sumner, he wrote, for all his harping on the supposed inexorability of nature's processes and the foolishness of trying to counter them, really did not know what he was talking about. A clergyman turned political economist, he lacked the knowledge of nature as it actually was that a training in science would have given him. But even so, Ward thought, the fallacies of his arguments should have been too patent to be overlooked:

> When a well-clothed philosopher on a bitter winter's night sits in a warm room well lighted for his purpose and writes on paper with pen and ink in the arbitrary characters of a highly developed language the statement that civilisation is the result of natural laws, and that man's duty is to let nature alone so that untrammeled it may work out a higher civilisation, he simply ignores every fact within the range of his faculties.[17]

Suppose, Ward continued, that we took the precepts of the nature-worshipers seriously and acted accordingly. The earth would then only support a much smaller population, at a miserable standard of living, if humankind renounced the unnatural use of inventions and interference and contented itself by gleaning the resources that nature unaided could

[15] Lester F. Ward, "Scientific Basis of Positive Political Economy" (orig. 1882), reprinted in *Glimpses of the Cosmos*, vol. 3, 34; "The Animal Population of the Globe" (1881), reprinted in *Glimpses of the Cosmos*, vol. 2, 358–359.

[16] Ward, "Scientific Basis of Positive Political Economy," 47–48; "The Sociological Meaning of Benevolent Institutions" (orig. 1884), reprinted in *Glimpses of the Cosmos*, vol. 3, 288; "The Gospel of Action" (orig. 1899), reprinted in *Glimpses of the Cosmos*, vol. 6, 59.

[17] Ward, "Professor Sumner's Social Classes," 304; "Mind as a Social Factor," 569.

provide. Instead, man "has, from the very dawn of his intelligence, been transforming the entire surface of the planet he inhabits," and largely for the better. The nature dear to the physiolaters was beneficially interfered with "every time a river is made navigable by dredging its channel." On their premises, "to dam a stream must be characterized as a 'vain' attempt to overcome a natural law."[18] And what could be more preposterous than to condemn such a useful action?

Ward acknowledged that people had done much damage to their environment, but he attributed it to the hitherto unplanned and natural, rather than planned and artificial, character of their social organization.[19] A society allowed to take its "natural" (uncoordinated, undirected) course, he reasoned, would manage nature as badly as it managed everything else. Ward always insisted that his social thought was intimately connected with and deeply nourished by his scientific work and his knowledge of the natural world, and indeed his case for planning and rationalizing the chaotic world of non-human nature duplicated his case for doing the same with the chaotic world of disorganized relations among human beings. He was the most articulate American critic in the post-Civil War years of the principle of laissez-faire most energetically espoused by Sumner.[20] To his eyes, the long persistence of irrational ways of life and the unregulated workings of a competitive free-for-all mimicked the failings and inefficiencies of nature in the purely physical realm and should be superseded, and for the same reasons, by a much greater degree of scientific government.[21] He was, by any measure one might use, the antithesis of a political or social conservative, uniting free-thought in religion, a feminism quite advanced for his time, and a deep distrust of tradition into an anti-laissez-faire progressivism that would win him the posthumous title of intellectual father of the welfare state and of the American

[18] Ward, "Mind as a Social Factor," 568, 569, 572; "The Political Ethics of Herbert Spencer" (orig. 1894), reprinted in *Glimpses of the Cosmos*, vol. 5 (New York, NY: G. P. Putnam's Sons, 1917), 63.

[19] E.g., Ward, *Dynamic Sociology*, vol. 1, 16, 73–74, 274; vol. 2, 88–89.

[20] Rafferty, *Apostle of Human Progress*, 117–119, 145–150.

[21] The argument is omnipresent in Ward's work; for an early and typical statement, see "Mind as a Social Factor," 571–573.

New Deal.[22] Unregulated economic competition he thought as wasteful as unregulated completion among species. Society, he asserted, was governed not too much but much too little. The state had acquired a bad name from the selfish abuse of its powers during most of the human past, Ward conceded, but from a democratically minded government, rightly understood, there was nothing to fear and everything to expect, for it was almost good by definition. It meant nothing more nor less than the application of intelligence to the improvement of human life, the opposite of standing back and letting things happen as they would.[23]

In 1884, at the same time that he was first setting out the essentials of his social philosophy, Ward offered a Promethean program for the earth in a concrete case, an extensive portion of the northern Great Plains widely considered too dry for cultivation and therefore for settlement. If the flow of the region's rivers were appropriated entirely for irrigation instead of being allowed to run uselessly away, Ward argued, the area could be reclaimed from its natural poverty and be "filled up by a thriving agricultural population." He dismissed as unimportant any collateral costs that might result from such a rearrangement of nature. The loss of navigation on the vanished rivers was the only one he bothered specifically to mention, and he thought it a trivial matter. That such a transformation of the upper Plains would be a feasible project in "this age of great engineering enterprises" Ward did not doubt, and he characteristically saw it as a project that only the government, not private capital, could and should undertake.[24]

Ward's later writings continued to dwell on and develop these themes. Writing in 1889 about "some social and economic paradoxes," he began by restating the one he had already made central to *Dynamic Sociology*: "the artificial is superior to the natural." He added another directed

---

[22] Henry Steele Commager, "Introduction," in Henry Steele Commager, ed., *Lester Ward and the Welfare State* (Indianapolis, IN: Bobbs-Merrill, 1967), xxii, xxxviii.

[23] This argument is also omnipresent; for some detailed statements, see Ward, "False Notions of Government," in *Glimpses of the Cosmos*, vol. 4 (New York, NY: G. P. Putnam's Sons, 1915), 64–71; "The Political Ethics of Herbert Spencer," 39–66; "Plutocracy and Paternalism" (orig. 1895), in *Glimpses of the Cosmos*, vol. 5, 231–240; and *The Psychic Factors of Civilization*, Chaps. 34–36 and 38.

[24] Lester F. Ward, "Irrigation in the Upper Missouri and Yellowstone Valleys," *Science* 4, #82 (1884), 166–68.

squarely at the Malthusian theory of population and its implied broader lesson of the narrow physical limits to human expansion and well-being. On the contrary, Ward wrote, "in industrial society, the means of subsistence increase more rapidly than population." The essential error of Malthus and his followers, he argued, had been to ignore the crucial role of invention in creating abundance: invention, he added, not as often misunderstood, the work of "a few great brains," but engendered by the whole of society's mental resources. The benefits of cooperation and of the productive use of intelligence increased with population to a degree that more than offset the greater demand for resources by more people.[25] Inventiveness, intelligence, and prosperity, Ward pointed out, all else being equal, were highest where people were most numerous and most densely clustered together.[26] Much more disposed to alter external than internal nature, he seized eagerly on Ellen Richards's term "euthenics."[27] Like her, he preferred it to eugenics as a means of enhancing human capacities and well-being. What chiefly limited the progress of mankind and of other species as well, Ward believed, was not any defects inherent to the organism, but the hostility of the environment. Remake the environment and the organism would thrive.

Ward's ideas put him squarely in the conservationist and not the preservationist camp. As Donald Worster has suggested, he was the philosopher of progressive conservationism no less than of the welfare state: "Ward's environmental outlook was thoroughly, even militantly anti-naturalistic"; he equated civilization with "the necessary, rational management of nature."[28] His particularly vivid awareness as a paleobotanist of the vast changes that the earth's surface had undergone may have made him less hesitant than most about proposing vast additional ones. If some of today's reasons for a natural scientist to think differently about nature

[25] Lester F. Ward, "Some Social and Economical Paradoxes," *American Anthropologist* 2, #2 (1889), 121, 125; "What Shall the Schools Teach?" (orig. 1888), in *Glimpses of the Cosmos*, vol. 4, 101.

[26] See, e.g., Ward, "Some Social and Economical Paradoxes," 125 and *Applied Sociology: A Treatise on the Conscious Improvement of Society by Society* (Boston, MA: Ginn and Company, 1906), 219–220, 227–232.

[27] Lester F. Ward, "Eugenics, Euthenics, and Eudemics" (orig. 1913), reprinted in *Glimpses of the Cosmos*, vol. 6, 390n.

[28] Worster, *Nature's Economy*, 174.

did not exist then, some did, and Ward was not altogether insensitive to them. Each of his few lapses into preservationism, though, was transparently a matter of personal special pleading. A botanist like himself, he acknowledged in 1881, did not see the landscape as others did. "Rich fields of corn are to him waste lands; cities are his abhorrence, and great areas under high cultivation he calls 'poor country;' while on the other hand the impassable forest delights his gaze, the rocky cliff charms him; thin-soiled barrens, boggy fens, and unreclaimable swamps and morasses are for him the finest lands in a State," but he granted that this set of preferences might well "seem as absurd to some as the witholding from tillage of great pleasure grounds in the form of hunting parks for the landed sporting gentry."[29] He agitated for two parks: one in Washington's Rock Creek in his own vicinity, where he enjoyed walking and botanizing, and another in Arizona's petrified forests, a rich source of fossil material for his own research, threatened at the turn of the century by the advent of a mill to grind the rock into industrial abrasives.[30] And although he ordinarily emphasized, against a prevailing folk belief in the rural origin of "great men," the vast intellectual benefits of growing up amid the opportunities and stimuli of a city, nostalgia for his own rural upbringing led him to argue, too, that children dwelling in cities needed far more opportunity than they usually had for close and frequent contact with nature.[31] He argued as well, though, that the love of nature was not a fundamental element of the human character but a relatively recent acquisition, one he saw no traces of in the literature of classical times.[32] On the whole, suggestions that nature untouched by human hands was better than nature improved are rare in his work. His usual preferences lay on the side of a thoroughly humanized earth: "in place of primeval forests we have enlightened populations; in place of wild beasts we have

[29] Lester F. Ward, "Field and Closet Notes on the Flora of Washington and Vicinity" (orig. 1881), in *Glimpses of the Cosmos*, vol. 2, 370–371.

[30] Ward, "Field and Closet Notes," 370; Lester F. Ward, "Report on the Petrified Forests of Arizona," (orig. 1900), reprinted in *Glimpses of the Cosmos*, vol. 6, 92–98.

[31] Lester F. Ward, "Early Education and Precocity" (orig. 1893), reprinted in *Glimpses of the Cosmos*, vol. 5, 29–30.

[32] Lester F. Ward, "Ethical Aspects of Social Science" (orig. 1896), reprinted in *Glimpses of the Cosmos*, vol. 5, 278–279.

statesmen and philosophers working out the problems of life, mind, and society."[33]

Ward developed his ideas within a community of like-minded natural scientists based in the nation's capital and associated with such public research institutions as the Smithsonian, the US Geological Survey, and the Bureau of Ethnology. The historian Michael Lacey has traced many of the intellectual roots of the early twentieth-century progressive conservation movement to this set of late nineteenth-century thinkers. In Lacey's words, they combined "a program of gradualist social transformation," one aimed particularly at expanding the role of egalitarian public action and narrowing the sphere of the market, with a conviction of "the duty of men to reconstruct and improve the earth."[34] A key colleague of Ward's in the Washington community was the geologist W J McGee (1853–1912), whom we have already encountered as the conservation movement's leading expert on water resources and whose devotion to efficiency made him insist on the omission of the wasteful periods after his first and middle initials. Though in his applied work he took water management for his particular province, he fitted it into a comprehensive Promethean philosophy of nature and society.

McGee, largely self-taught like Ward himself and many of their generation of earth scientists, advanced from the routine of mapping and classifying to questions of glacial geology and tectonics after joining the recently founded USGS in 1881. Before long, he expanded his interests even further into anthropology (under its then-common label of ethnology) and the development, both past and future, of human civilization. More than Ward or other members of the group, he proposed concrete applications of their speculations about the fruitful application of mind to nature. He was one of the most active members of the progressive

[33] Lester F. Ward, "The Immortality That Science Teaches" (orig. 1887), reprinted in *Glimpses of the Cosmos*, vol. 4, 62.

[34] Michael Lacey, "The Mysteries of Earth-Making Dissolve: A Study of Washington's Intellectual Community and the Origins of American Environmentalism in the Late Nineteenth Century," Ph.D. dissertation, George Washington University, 1979, 4, 284.

conservation movement during the heyday of White House patronage it enjoyed in the last years of Theodore Roosevelt's presidency.[35]

As had Saint-Simon's, McGee's belief in progress took the form of a theory of stages.[36] Though the two theories were far from identical, they agreed on some important points, among them their detachment from orthodox Christian conceptions. (A scandalized Samuel Pierpont Langley, secretary of the Smithsonian, described both Ward and McGee, correctly enough, as atheists and radicals, and he sought to curb their access to government-sponsored outlets for publishing some of their more controversial views.[37]) McGee also shared Saint-Simon's admiration of Bacon, whose goal of making humankind "the master of nature" and whose Promethean rationale for the pursuit of science—"the course of nature has come to be investigated in order that it may be re-directed along lines contributing to human welfare"—he endorsed.[38] Above all, he and the French technocrat concurred in seeing the rising power of the human mind to subjugate and remake nature as the most fundamental characteristic of modern social evolution. Henceforth, they expected, the rationality chiefly typical so far of the practice of the natural sciences would be applied to all spheres of life on earth, human and nonhuman. Intelligence and planning would put an end to chaos and disorder.

Two tasks faced human beings, McGee wrote: in their relations among themselves, increasing "the efficiency of the State," and in their relations with nature, doing the same with the soils and the species on whose hitherto blind, unguided striving all of the earth's evolution had previously rested, "transforming the face of nature, by making all things better than they were before, by aiding the good and destroying the bad among animals and plants," obliterating the harmful and useless and multiplying the valuable. Humankind's success already in doing so could be measured by the fact that:

[35] On McGee's career and thought, see Whitney R. Cross, "W J McGee and the Idea of Conservation," *The Historian* 15, #2 (1953), 148–162 and Lacey, "The Mysteries of Earth-Making Dissolve," Chap. 4, "W J McGee and the Apotheosis of Nature," 284–342.

[36] Lacey, "The Mysteries of Earth-Making Dissolve," 297–300.

[37] Cross, "W J McGee," 152.

[38] W J McGee, "The Trend of Human Progress," *American Anthropologist* n.s. 1, #3 (1899), 444; "Water as a Resource," 37.

> About one-tenth of all the lands of the lands of the earth are now appropriated to human use for fields and domiciles, factories, and highways, and no plants or animals are permitted to remain thereon that do not subserve human needs; some three-tenths more of the lands of the earth, with all things pertaining thereto, are more or less perfectly appropriated to human use; and almost all of the remainder, save the mountain tops and the ice-burdened poles, have become tributary to human needs and pleasures.[39]

McGee, like the other conservationists, did not ignore the environmental damage that human action—particularly unregulated private action—had done. He called attention frequently to the signs of soil erosion from the land, of the siltation of rivers, of larger floods alternating with longer periods of low streamflow, and of water pollution by industry, cities, and agriculture. But he ascribed them not to some general unwisdom of human tampering with nature, but rather to human tampering badly directed, driven by greed and ignorance rather than by science. The best solutions to these problems, to his mind, involved "more intensive transformation of nature rather than less," as Lacey summed them up, particularly the idea McGee made most central to his work in conservation late in life, the comprehensive management of entire drainage basins and river systems for the multiple purposes they could be made to serve.[40]

By imposing reason and purpose on the inefficient workings of nature, humankind would make it a far more bountiful reservoir of resources. McGee looked forward to the day when the Pacific Ocean, as vast a field for effort as the earth could offer, would be converted from a marine wasteland into "feeding grounds for useful organisms, just as are the narrower fields and pastures of the land."[41] Though, like other conservationists, he argued for the comprehensive development of rivers in order to

[39] W J McGee, "The Relations among the Resources," in *Addresses and Proceedings of the National Conservation Congress Held at Seattle, Washington, August 26–28, 1909* (Washington, DC: National Conservation Congress, 1909), 99; W J McGee, *The Earth the Home of Man*, Anthropological Society of Washington Special Paper 2 (Washington, DC, 1894), 24–25, 28.

[40] See, e.g., W J McGee, "Our Inland Waterways," *Popular Science Monthly* 72, #4 (1908), 302–303; Emma R. McGee, *Life of W J McGee … with Extracts from Addresses and Writings* (Farley, IA: privately printed, 1915), 97, 188–89, 192; Lacey, "The Mysteries of Earth-Making Dissolve," 339.

[41] W J McGee, "Problems of the Pacific: The Great Ocean in World Growth," *National Geographic Magazine* 13, #9 (1902), 337; McGee, *The Earth the Home of Man*, 26.

extract their full potential for power generation, for McGee, at least, it was not because of any haunting fear that fossil energy would run short. Seeing as "indefinitely remote" the day when the supply of usable hydrocarbon fuels from the earth would be exhausted, especially as new methods of efficient extraction were developed, he felt no serious fears for the availability of the inanimate forces "by which the wheels of human progress will be kept in motion."[42] He saw no physical obstacle to a future USA supporting a population of a billion people on its own land and water resources with existing technologies of food production. Intensive agriculture would, he granted, mean a great alteration of the face of nature, especially in the more densely populated East, but an alteration of which he approved; it was "the true way to conquest over nature, to the subjugation of natural forces for human welfare—for the face of nature must be transformed." It would bestow on the fortunate eastern seaboard "landscapes more completely artificialized than those between London and Liverpool or Paris and Amiens," filled with an upstanding population of sturdy and independent tillers of the soil. Moreover, McGee added, echoing Percy Bysshe Shelley's ruminations as an Oxford student, one billion was the limit of population within the USA only until human ingenuity mastered the secret of manufacturing water (in his view, the key limiting resource): "then will mankind rise to a new plane of progress, and the desert will blossom."[43]

All of the environmental progress that McGee envisioned depended on the assertion or reassertion of control by the government, standing for the common good, over the earth and its essential resources. River development on the scale that he proposed was plainly incompatible, as he liked to point out, with the unlimited rights of private property, as were his plans for the land. McGee deplored the way in which the nineteenth-century federal and state governments had recklessly disposed of vast acreages of the public domain. They had plainly failed, he thought, in their commendably egalitarian purpose, which had been the Jeffersonian one of establishing the bulk of the population on a basis of economic, and therefore political and social, independence. Concentrated private

[42] W J McGee, "The World's Supply of Fuel," *Forum* 7 (1889), 566.

[43] W J McGee, "How One Billion of Us Can Be Fed," *The World's Work* 23, #4, 449, 451.

ownership and control had become the rule instead, relegating the majority of the people to "conditions of industrial dependence akin to those of the yeomanry and peasantry in days of feudalism." As Lacey sums the matter up, McGee and his associates linked their environmental Prometheanism to a reversal of the trend of privatization and a reinvigoration of a scientifically managed public domain.[44]

McGee was one of a number of earth scientists of this time to suggest that the appearance of intelligent humankind wielding the fruit of its intelligence, technology, was opening a new era in the planet's physical and biological evolution comparable to the major epochs of the geological past. He himself labeled the change the emergence of the "psychosphere," which now represented a major physical dimension of the planet just as did the much older ones that had successively emerged: the nucleosphere (the still molten core), lithosphere (solid rock), atmosphere (gases), hydrosphere (water), phytosphere (plants), and zoosphere (animals). With its development, human intelligence "seems to be enveloping our planet and commencing the control of the rock-sphere, the water-sphere, and even the air-sphere for the good of humanity," representing the force of conscious beings "united in thought and purpose" toward "the unceasing conquest of nature." Ward likewise identified the emergence of thought as beginning a new great epoch of the earth's physical development.[45]

Of the Americans of their day who wrote along similar lines—others included the geologists James Dwight Dana and Joseph Le Conte—Ward and McGee were the most vocally enthusiastic about what the change had in store. In very recent years, the term "Anthropocene" has come into wide use as a name for a new era of earth history dominated by the forces of human activity. Its chief predecessor was one given the most currency by two other leading earth scientists in the early twentieth century, the Russian geochemist Vladimir Ivanovich Vernadsky (1863–1945) and the French paleontologist and Jesuit priest Pierre Teilhard de Chardin (1881–1955). They described the emergence of a "noosphere," a realm

[44] McGee, *Life of W J McGee*, 94; Lacey, "The Mysteries of Earth-Making Dissolve," Chap. 5, "The Conservation Movement and the Restoration of the Commons," 343–416.

[45] W J McGee, "The Geospheres," *National Geographic Magazine* 9, #10 (1898), 447; Lester Frank Ward, "The Course of Biologic Evolution" (orig. 1890), reprinted in *Glimpses of the Cosmos*, vol. 4, 218–219.

of organized human thought, as a transforming event in the planet's evolution, one equivalent in importance to the previous appearance of the rock shell of the geosphere or lithosphere and the living envelope of the biosphere on its surface. In the works of both men and of their leading disciples, the concept of the noosphere carried a tinge of political progressivism—as indeed it was hardly compatible with conservatism in any meaningful sense of the word. Active in the liberal opposition before 1917, Vernadsky voluntarily remained in Russia after the Revolution to continue his work. The more restrained of the two writers, he largely confined himself to documenting the physical and chemical evidence for the earth's transformation, but he did not hide his belief that it was both inevitable and desirable.[46] Teilhard, the unorthodoxy of whose views caused his superiors in the Catholic hierarchy to withhold approval for their publication, was more explicit. "In becoming planetised," he wrote in a typical vein in 1947, "humanity is acquiring new physical powers which will enable it to super-organise matter." "Who can say," he asked, "what forces may be released, what radiations, what new arrangements never hitherto attempted by Nature, what formidable powers we may henceforth be able to use, for the first time in the history of the world?" A modern theologian much influenced by Teilhard nonetheless felt compelled to dissent from the Frenchman's insistent environmental Prometheanism, from his "metaphors of conflict, confrontation, attack and conquest" casting human beings as adversaries triumphing over nature; Teilhard paid "inadequate attention to the integral functioning of the natural world" and showed "no awareness of the increasing desolation of the Earth" produced by the reckless use of technology.[47]

At the other extreme from this cosmic level of generalization, a discipline of geography took shape during the nineteenth century as a detailed

[46] Vladimir I. Vernadsky, "The Transition from the Biosphere to the Noosphere" (1938), trans. William Jones, *21st Century* (Spring-Summer 2012), 16–31 and "Some Words About the Noosphere" (1943), trans. Rachel Douglas, *21st Century* (Spring 2005), 16–21. On Vernadsky's career, see Kendall E. Bailes, *Science and Russian Culture in an Age of Revolutions: V. I. Vernadsky and His Scientific School, 1863–1945* (Bloomington, IN: Indiana University Press, 1990).

[47] Pierre Teilhard de Chardin, *The Future of Man*, trans. Norman Denny (New York, NY: Harper & Row, 1964), 183; Thomas Berry, "Teilhard in the Ecological Age," in Arthur Fabel and Donald S. John, eds., *Teilhard in the 21st Century: The Emerging Spirit of Earth* (Maryknoll, NY: Orbis Books, 2003), 57–73 (quotations from 59, 66, 71).

science of the earth's surface and of the interactions between it and its human occupants. Western (though American and British more than French, and especially more than Russian) geographers tended in this period to interpret human societies as products of the physical environment, often to politically conservative ends. Several important figures, though, disputed this environmental determinism by calling attention to the opposite direction of influence, the human role in changing the environment. Two of them stand out for the way they combined a marked political radicalism with strongly Promethean prepossessions about the benefits of human-induced change.

One of them was the Frenchman Élisée Reclus, whose turbulent career belied his mild and engaging personality. Converted to socialism in the 1860s, he dropped his geographical work to support the Paris Commune in 1871. He then continued his research and writing in forced exile in Switzerland and later Belgium while emerging as one of Europe's foremost spokesmen for anarchism and for such related progressive causes as religious free-thought, the equality of the sexes, and animal rights.[48] His overall philosophy of nature and society was a tempered and benevolent Prometheanism. An older contemporary, the American geographer George Perkins Marsh, in 1873 singled out just this tendency of Reclus's work for comment. Marsh's own volume *Man and Nature* (1864) was a landmark in the development of environmental awareness.[49] It examined the effects, deliberate and especially inadvertent, of human action on the earth's surface, acknowledging the many beneficial results, but focusing on the harmful and destructive ones. Issuing a revised version a decade later, Marsh took note of Reclus's two-volume *L'Homme et la terre*, which had appeared in the interim, calling it an "admirable work" and on one important point, "a complement to my own." For Reclus, he observed, had, without ignoring the harms, emphasized the improvements upon

[48] On Reclus's career and thought, see Gary Dunbar, *Élisee Reclus. Historian of Nature* (Hamden, CT: Archon Books, 1978) and John P. Clark and Camille Martin, eds., *Anarchy, Geography, Modernity: Selected Writings of Élisee Reclus* (Oakland, CA: OM Press, 2013). Clark and Martin emphasize the preservationist elements in Reclus's thought more than I do, but acknowledge the existence of his Promethean side as well.

[49] David Lowenthal, *George Perkins Marsh: Prophet of Conservation* (Seattle, WA: University of Washington Press, 2000).

nature produced by human action, drawing "an attractive and encouraging picture of the ameliorating influence of the action of man," whereas he himself had done the opposite.[50] The American businessman, legislator, and diplomat stressed the dangers of human modification of the earth's surface; the radical European exile, its benefits. Reclus, with equal courtesy, had drawn the same distinction between their points of view in 1864 when he reviewed Marsh's volume for the influential French periodical *La Revue des deux mondes*. Acknowledging the value of its warnings, Reclus thought it necessary to balance them by reminding readers in detail of "what can be achieved by the persevering will of man" at its best: the drainage of swamps and marshes and the restoration of their inhabitants to health, the greening of deserts, the afforestation of wastelands, and even the enhancement of nature's scenic beauties by the application of human art. In closing, he hoped that the future would bring the atmosphere and rainfall under human control as the land surface was already.[51]

L'homme et la terre and Reclus's two popular books on streams and mountains indeed largely celebrated the improvement of nature in the way that Marsh had noticed. Reclus dwelt enthusiastically on the way in which the persevering labors of the masses in clearing, fertilization, terracing, irrigation, and drainage (sometimes even of whole lakes or inlets of the sea) had created the world's fertile farmland. To facilitate movement, they had bridged rivers for land traffic and made them navigable for ships, pierced mountains with tunnels, improved harbors, and broken through one of the earth's great isthmuses, with the other soon to follow.[52] On mountains as elsewhere, Reclus observed approvingly, "man creates a new earth adapted to his wants," if need be removing the mountains themselves to clear obstacles from his path.[53] No longer superstitiously worshiping rivers, human beings now made the best use of them for irrigation, industrial power, navigation, water supply, and waste

---

[50] George P. Marsh, *The Earth as Modified by Human Action* (New York, NY: Scribner, Armstrong & Co., 1874), viii.

[51] É. Reclus, "De l'action humaine sur la géographie physique: L'homme et la nature," *Revue des deux mondes* 54 (1864), 762–771 (quotation from 768).

[52] "Le travail de l'homme," in É. Reclus, *La Terre: Description des phénomènes de la vie du globe: II: L'océan—L'atmosphère—La vie* (Paris: Hachette, 1869), 670–757.

[53] É. Reclus, *Histoire d'une montagne* (Paris: J. Hetzel, 1880), 244–246 (quotation from 245).

disposal while curbing the destructiveness of their floods and eliminating their "pestilential morasses."[54] Human action had indeed destroyed much plant and animal life, Reclus observed, but he confidently expected it in the future to shift from impoverishing to enriching the earth's biota.[55] Ambivalent at times about human engineering, sympathetic to preservationists who feared too great a loss of untouched nature and the psychological benefits it could bring, harshly critical of changes that defaced the earth's surface, Reclus nonetheless asserted that not only in its useful output, but even in its scenic beauty, a land enhanced by sensitive artifice and design was superior to one made by nature alone. Some regions of dense agricultural settlement and intensive cultivation, such as Lombardy and Flanders, he thought, best exemplified the ability of farmers to increase the land's productivity and its attractiveness at the same time. If science and society progressed together, but only if they did, the planet would be transformed from what it had originally been into "an immense organism working tirelessly for humankind's benefit" and, embellished and beautified by art, "the garden dreamt of by the poets of all the ages."[56]

Reclus's fellow geographer, the Russian exile Petr Alekseevich Kropotkin (1842–1921), likewise united a radical political activism with a vision of nature improved by human action, and, indeed, he expressed few of the Frenchman's occasional reservations about the beneficence of the latter. Born into the Russian nobility, Kropotkin, while doing his military service in the Russian Far East, interested himself in the study of the earth's surface. He made important contributions to the chronology of the ice ages and the understanding of the distribution of plant and animal species on the basis of research expeditions in Siberia and Manchuria. In the same years, his growing political radicalism led him into illegal revolutionary circles and eventually to a tsarist prison. He escaped in 1876 to spend decades of exile (and a few more spells of imprisonment) in western Europe as a geographical scholar and a prophet of anarchism.[57]

[54] É. Reclus, *Histoire d'un ruisseau* (Paris: Hachette, 1869), 17.

[55] Reclus, "Le travail de l'homme," 736–745.

[56] É. Reclus, "De la sentiment de la nature dans les sociétés humaines," *Revue des deux mondes* 63, #2 (15 May 1866), 379; Reclus, "De l'action humaine," 771; Reclus, "Le travail de l'homme," 757.

[57] On his life, see Martin A. Miller, *Kropotkin* (Chicago, IL: University of Chicago Press, 1976).

The anarchism that Kropotkin preached was not one of libertarian laissez-faire. Challenging the centralized power of the sovereign nation-state as too likely to be abused for the benefit of the wealthy and influential, it did not substitute the sovereignty of the private individual dear to orthodox economics. It entailed local communal and collective rather than individual ownership of the key resources of the land. Kropotkin thought the isolated, asocial model of man an illusion and a pernicious one, deliberately fostered in Western capitalist society to justify the grabbing by the few—whose outstanding personal qualities had supposedly earned them the privilege—of the resources that had been created by the many. Vast and otherwise indefensible inequalities of wealth were the result. Most of the technologies of production, Kropotkin maintained, originated in society and not in the work of isolated individuals whose genius entitled them to outsized shares of the total product.[58] He denied that competition, either in the animal or in the human world, generated wealth and advanced human well-being so effectively as cooperation, disputing a Social Darwinist thesis he saw as inconsistent with the facts of both biogeography and social science.

He drew on his research in the harsh climates of northeastern Russia to dismiss the idea that, in the face of potential overpopulation, members of a given species incessantly fought one another for the meager subsistence available to them. He saw underpopulation, rather, in the north, an absence of the numbers that could potentially thrive; they were kept down not by struggle within the species but by such natural extremes as cold and drought. Those species flourished best in this setting that cooperated, in what Kropotkin called "mutual aid," to ensure their collective survival in ways that isolated individuals could not. The lesson, in his hands, led straight to a corollary for human beings: a society whose members competed would be poorer than one whose members cooperated. Cooperated against what? Against their chief adversary, the harshness of their natural environment.[59] Kropotkin praised nature in one sense of

[58] See, e.g., Peter Kropotkin, *The Conquest of Bread and Other Writings*, ed. Marshall S. Shatz (Cambridge: Cambridge University Press, 1995), 19–20, 152–157.

[59] For a summary and analysis, see Daniel P. Todes, *Darwin without Malthus: The Struggle for Existence in Russian Evolutionary Thought* (New York, NY: Oxford University Press, 1989), ch. 7, "Kropotkin's Theory of Mutual Aid," 123–142.

the word, denying as he did that nature's processes of evolution worked in unconscionable ways and rewarded brutal selfishness, but at the same time his argument implied the desirability of bringing the biophysical environment under human control.

And Kropotkin explicitly celebrated the way in which "the human race … had learned to subdue the forces of nature." Its earliest great adversary was "Nature, vast, unknown and terrific, with whom they had to fight for their wretched existence." Modern man had emerged victorious, having "cleared the land, dried the marshes, hewn down forests, made roads, pierced mountains." How, Kropotkin asked, did one know a country to be civilized? "The forests which once covered it have been cleared, the marshes drained, the climate improved. It has been made habitable. The soil, which bore formerly only a coarse vegetation, is covered today with rich harvests. The rock walls in the valleys are laid out in terraces and covered with vines. The wild plants, which yielded nought but acrid berries, or uneatable roots, have been transformed by generations of culture into succulent vegetables or trees covered with delicious fruits. Thousands of highways and railroads furrow the earth, and pierce the mountains. … The rivers have been made navigable … artificial harbours, laboriously dug out and protected against the fury of the sea, afford shelter to the ships. Deep shafts have been sunk in the rocks; labyrinths of underground galleries have been dug out where coal may be raised or minerals extracted."[60]

Human beings, far more than any other animals, had acquired the power not merely to defend themselves against nature, but to remake it. As a result, they had created wealth in unprecedented quantities. But if a competitive system was the ideal form for organizing human energy, Kropotkin argued, there was no way of explaining the gross disparity between the high degree of mastery over nature that had been achieved and the meager degree to which the majority of the population had benefited, even in the wealthiest countries of Europe. "In our civilized societies we are rich. Why then are the many poor?"[61] The disparity between the possible and the actual he found most glaring in farming, the livelihood

[60] Kropotkin, *The Conquest of Bread*, 11, 13.

[61] Ibid., 12–13.

of the majority of the world's population and the support of all. With poverty and hunger so common, the cry of agricultural overproduction, one heard frequently in Europe and North America, must be mistaken. It seemed indeed to be borne out by widespread land abandonment, farms "going to ruin," their laboriously wrought improvements allowed to decay as they reverted to nature.[62] But the trend, Kropotkin asserted, grew not out of an absolute excess of output, but rather out of the inadequate purchasing power of the many and out of a harmful economic system that drove rural populations to factories in the cities. What was needed was not to restrict output further, but, rather, to maintain and enlarge the extent to which humankind had replaced nature with cultivation.[63] As it was, an "appalling" quantity of land, even near such large cities as London, had been taken out of production and turned back to a more natural state merely to make preserves for pheasant and deer hunting for the aristocracy. Kropotkin likewise deplored the way in which nobles and the clergy had appropriated land in pre-Revolutionary France and allowed it to revert to nature. Conversely, he praised his favored form of social organization, the village community, for its particular ability to achieve and maintain human mastery over the land. In France, he wrote, "Canals were maintained, forests were cleared, trees were planted, and marshes were drained by the village communities from time immemorial; and the same continues still."[64]

The soil itself, Kropotkin insisted, was much less a gift of nature than a creation of human effort. The most productive lands in the world, he asserted with many examples, were not those that had the richest natural soils, but precisely the ones that had the poorest, peat bogs and sandy wastes. They were the ones whose human inhabitants could not rely on the gifts of nature, but had been obliged to manipulate and improve the ground beneath their feet and had thus brought it to a height of

[62] P. Kropotkin, *Fields, Factories and Workshops*, revised edition (New York, NY and London: G. P. Putnam's Sons, 1913), 93.

[63] Ibid., 64–65; Kropotkin, *The Conquest of Bread*, 17–18.

[64] Kropotkin, *Fields, Factories and Workshops*, 95–96; P. A. Kropotkin, *Mutual Aid: A Factor of Evolution* (New York: McClure Phillips & Co., 1902), 230–231, 231n, 245 (quotation). For examples from other countries, see also 154–155, 249, 256, 258–259.

development unknown elsewhere.[65] As a rule, Kropotkin asserted, "the soil is always *made*." The clearing of "a few acres of virgin soil" or the building of "roads, bridges, canals" enriched future generations unless the improvements were allowed to decay and the land lapse back into a natural state.[66] He was as relentless a critic of what he called "the overpopulation fallacy" as he was of the thesis of agricultural overproduction.[67] If there were too many people for the amount of food produced, it was because the production even of food had been, and necessarily was, deliberately restrained to keep up prices under an economy of market exchange and private ownership. One could never speak of overpopulation; it was impossible to calculate the number of people a given area of land could support because of the increased yields that future inventions must bring, and because of the meager use currently being made even of many ones already available. One could only guess at the eventual output possible from cropland "covered with glass ... ploughed by steam, improved by manures, or enriched by artificial soil obtained by the pulverization of rocks," to say nothing of what advances the future might make in the application of new forms of energy and of biotechnology.[68] But making the earth as productive and its people as well-fed as they could be required a political and economic revolution. Once "a socialistic organization of work will be established instead of the present capitalistic one," "as soon as the peasant realizes that he is no longer forced to support the idle rich by his toil, [n]ew tracts of land will be cleared, new and improved machines set agoing."[69]

Another earth scientist, the twentieth-century American geologist Kirtley F. Mather (1888–1978), took an equally Promethean view both of the human power to alter the environment for the better and of the amplitude of the planet's resources to meet human needs. Mather coupled his environmental beliefs with a social and political progressivism that stopped well short of Communism—he was a strong critic of the

[65] E.g., Kropotkin, *Fields, Factories and Workshops*, 76, 92, 110, 121.
[66] Ibid., 126; Kropotkin, *The Conquest of Bread*, 162–163.
[67] Kropotkin, *Fields, Factories and Workshops*, 136–137, 158–161, 233.
[68] Kropotkin, *The Conquest of Bread*, 72, 192–193, 195–196.
[69] Ibid., 180.

Soviet Union's human rights record—but went far enough to the left to make him, as a member of the Harvard University faculty, one of the principal academic targets of American red-baiting from the 1930s through the 1950s.[70] As attached to the cause of environmental as of social progress, Mather drew upon his professional expertise to assess favorably the prospects for human beings as inhabitants of their planet. Species disappeared, he wrote, when their environment changed too rapidly or too drastically for them to adapt in time. Humankind alone was exempt from such a fate if it acted wisely. "Man has placed himself in control of external conditions to an extent immeasurably greater than has any other creature. He has practically drawn the teeth of environment." Not only did technology make human life possible in any setting on the earth, it could change the settings themselves. Humankind "drains the swamp, irrigates the desert, tunnels the mountain, bridges the river, digs the canal, conditions the air in home, factory and office." Nor need there be any fear that the natural resources from which the tools of control were fashioned would run short. Technology had made and was continuing to make agriculture ever more productive. Substitutes could be found or made and efficiencies of use improved for key mineral resources.[71] The supply of petroleum might indeed give out, but alternative hydrocarbons in the earth's crust that future innovations in extraction would make available were "almost unbelievably abundant," to say nothing of the potential of renewable energy and of nuclear fission.[72]

Mather's 1944 book *Enough and to Spare* was the decade's most authoritative affirmation of natural resource abundance under human management. "The gloomy prediction of Malthus," it asserted, "does not now apply and, if present trends continue, never will apply to man."[73] In a review in the independent Marxist journal *Science and Society* in 1949,

[70] Kennard B. Bork, "Kirtley Fletcher Mather's Life in Science and Society," *Ohio Journal of Science* 82, #3 (1982), 74–95.

[71] Kirtley F. Mather, "The Future of Man as an Inhabitant of the Earth," *Scientific Monthly* 50, #3 (1940), 193–203 (quotations from 196, 197).

[72] Kirtley F. Mather, "Petroleum—Today and Tomorrow," *Science* n.s. 106, #2764 (1947), 603–609 (quotation from 609).

[73] Kirtley F. Mather, *Enough and to Spare: Mother Earth Can Nourish Every Man in Freedom* (New York, NY: Harper & Brothers, 194), 70.

Mather described William Vogt's influential neo-Malthusian manifesto *The Road to Survival* as "essentially a bad book." By interpreting scarcity as a fact of nature, Vogt ignored, in Mather's view, all of the technological and social processes by which human society could, if it chose, forestall impending shortages of resources and make them again plentiful. "It is man's failure to utilize wisely and distribute justly the resources of the earth, rather than the inherent limits of those resources, that poses the real problem." What was needed was a proper organization of world society along democratic and egalitarian lines.[74] What was also needed, Mather wrote in *Enough and to Spare*, was an acceptance of planning guided by the best scientific knowledge and a realization, not that "the best government is the one that governs least," but that it was "the one that provides most adequately for the safety, health and physical welfare of all its citizens." Should these changes occur, no environmental threats to human welfare need be of serious concern.[75] Mather, too, echoed Saint-Simon in a statement that might almost have served as a Promethean password: the "golden age," he wrote, "if any, is in the future rather than in the past," and he summoned "everyone who believes in liberty, equality and fraternity" to help bring it about.[76]

Prometheanism took different but complementary forms in the more basic sciences of chemistry and physics. The French chemist Pierre Eugène Marcellin Berthelot (1827–1907) did as much as anyone in the Europe of his time to exemplify the linkages of first-rate natural science, outspoken political progressivism, and the improvement of nature. As a member of the Senate and of the French Academy as well as of the Academy of Sciences, he fought for such causes as antimilitarism and a secular system of education and against clericalism, class distinctions, and the hold of tradition on political and cultural life, which he associated with the dark forces of irrationality and obscurantism. His scientific work had close connections with his political thought. Through it he sought to deal a fatal blow to a vitalist chemistry whose insistence on the uniqueness and

[74] Kirtley F. Mather, review of William Vogt, *Road to Survival*, *Science and Society* 13, #2 (1949), 170–171.

[75] Mather, *Enough and to Spare*, 140.

[76] Mather, "The Future of Man," 202; *Enough and to Spare*, 152.

irreducibility of living processes he considered a prop of reactionary mysticism. Two of his principal research programs directed toward this goal involved the synthesis of a wide range of organic compounds by purely physical means in the laboratory and the demonstration that the amount of heat released by the combustion of substances within the body did not differ measurably from that released by their combustion outside it.[77]

Berthelot saw scientific and social progress as inseparable, and his Promethean expectations centered on the ever-expanding control and enrichment of nature and the creation of permanent material abundance. As Shelley had done, he looked forward to the replacement of agriculture by the far more efficient industrial manufacture of food. No longer needed for farming, the earth could then be made "a vast garden, irrigated by the flow of groundwater," its landscape designed for beauty and pleasure, and offering "the abundance and the joys of the legendary golden age."[78] There need be no fear that energy resources would ever fall short of human demands, Berthelot maintained. Though fossil fuels might be used up, science would furnish cleaner, safer, and perpetually abundant substitutes by harnessing the potential of running water and of the heat of the sun and the earth's core.[79]

So, too, though human demand might deplete the mineral resources that the earth afforded, creative chemical synthesis would respond with "an inexhaustible multitude of new substances, similar and superior to the natural ones."[80] Berthelot prided himself, indeed, on having helped to make chemistry something other and better than a mere natural science.

[77] On Berthelot's life and scientific program, see M.P. Crosland, "Berthelot, Pierre Eugène Marcellin," in Charles Coulston Gillispie, ed., *Dictionary of Scientific Biography*, vol. 2 (New York, NY: Charles Scribner's Sons, 1970), 63–72 and John Hedley Brooke, "Overtaking Nature: The Changing Scope of Organic Chemistry in the Nineteenth Century," in Bernadette Bensaude-Vincent and William R. Newman, eds., *The Artificial and the Natural: An Evolving Polarity* (Cambridge, MA: MIT Press, 2007), 275–292. On his politics, see also Robert Fox, *The Savant and the State: Science and Cultural Politics in Nineteenth-Century France* (Baltimore, MD: Johns Hopkins University Press, 2012), 264–266.

[78] See, e.g., M. Berthelot, *Science et morale* (Paris: Calmann Lévy, 1897), 8, 121–122; M. Berthelot, *Science et libre pensée*, 2nd edition (Paris: Calmann-Lévy, 1905), 26, 406, Berthelot, *Science et morale*, 513–514; *Science et libre pensée*, 64, 184–193.

[79] M. Berthelot, *Science et education* (Paris: Société française d'imprimerie et de librairie, 1901), 18, 376; *Science et morale*, 511–512.

[80] Berthelot, *Science et education*, 389.

"Chemistry," he declared in an often-quoted phrase, "creates its own object." His esteem for the artificial and the synthetic went along with frequent expressions of contempt for the natural. He gloried in the ability of the laboratory scientist to outdo the meager array of substances nature had managed to create and to invent new synthetic materials of greater usefulness to humankind.[81] He likewise found the natural world cruel and immoral and as greedy in its depredations on human productions as it was slow and inefficient in its own. He respected it chiefly where it most closely approximated human life, especially in the coordination and intelligent purpose displayed by such social animals as ants and beavers.[82]

The writings and activities of a cluster of more aggressively left-wing figures in Great Britain in the 1920s and 1930s did much to create the impression that science and socialism went naturally together.[83] The crystallographer J.D. Bernal (1901–1971) was at once the most consistently radical of the group in his politics and the most outspoken advocate of an environmental Prometheanism that he saw as its corollary. Born and raised in rural Ireland, Bernal went to England for school and continued on to Cambridge shortly after the end of the First World War, studying physics and chemistry while gradually replacing the Catholicism of his upbringing with Marxian socialism. In a distinguished scientific career focused on the use of X-ray crystallography to determine the structure of biological compounds, he found time to engage in much political work at home and internationally as a Communist—who remained loyal to the Party even through the events of 1939 and 1956—and to write prolifically on the relations of science and society.[84]

His achievements in natural science did not make Bernal an admirer of nature, which he regarded as blind and inefficient. "The cardinal tendency of progress," he wrote, "is the replacement of an indifferent chance

[81] Marcellin Berthelot, *Chimie organique fondeé sur la synthèse*, vol. 2 (Paris: Mallet-Bachelier, 1860), 811 (quotation); Berthelot, *Science et education*, 389 (quotation); see also, e.g., M. Berthelot, *Science et philosophie* (Paris: Calmann Lévy, 1886), 33, 57, 61–62, 91–92; *Science et libre pensée*, 64, 190.

[82] Berthelot, *Science et libre pensée*, 371–375; *Science et philosophie*, 172–175.

[83] Gary Werskey, *The Visible College: The Collective Biography of British Scientific Socialists of the 1930s* (New York, NY: Holt, Rinehart & Winston, 1979).

[84] The standard biography is Andrew Brown, *J. D. Bernal: The Sage of Science* (Oxford: Oxford University Press, 2005).

environment by a deliberately created one." Human intelligence would fashion "a new artificial Nature much more complicated, and at the same time much more fluid and amenable to human creative genius than ever Mother Nature provides." He was as indisposed to look to nature for moral guidance as for skilled workmanship or practical wisdom: "Goodness is purely social; there is no such thing as natural goodness." Nature was only valuable and useful insofar as it is "perceived and worked upon by man."[85] The ideas he expressed in the 1930s about how to work upon it had a generous scope. The application of science, he thought, could vastly increase the cultivable area of the earth by reclaiming deserts and by farming the oceans. It could develop the means of extracting mineral resources where they occurred in trace concentrations and of inventing more efficient synthetic substances to do their work, a prospect that engaged him as much as it had Berthelot. Future civil engineers would be occupied in "moulding the earth's surface … and in changing the climate." "It will no longer be a question of adapting man to the world but the world to man." A world fully adapted to that purpose would furnish enough food and other resources to support "a population thousands or millions of times that which occurs at present on our globe."[86]

When the next decade unveiled the means of liberating the energy of the atom, Bernal's hopes for what people could do with the earth rose still further. He campaigned vigorously against the military use of nuclear power, but at the same time wrote with enthusiasm about its possible peaceful applications in environmental engineering. Already, he wrote in 1945, it "can be used to dig canals, to break open mountain chains, to melt the ice barriers and generally to tidy up the awkward parts of the world."[87] Already, too, he maintained, such an abundant new source of energy gave the lie to the supposed physical scarcity of essential natural resources, making it clearer than ever that "the only limit to human capacity is to be found in society and not in nature." With its conversion to peaceful use, he wrote in the 1950s: "Man would truly become

[85] J.D. Bernal, *The Freedom of Necessity* (London: Routledge & Kegan Paul, 1949), 60–61, 81; *World Without War* (London: Routledge & Kegan Paul, 1958), 266.

[86] J.D. Bernal, *The Social Function of Science* (New York, NY: Macmillan, 1939), 348, 367, 372, 370–380.

[87] J.D. Bernal, "Everybody's Atom," *The Nation* 161, #9 (1 September 1945), 203.

master of the world in the material sense. With gigantic, atom-powered machines he will be able to build on it, mould its surface and burrow into its depths at will." Molding the earth's surface would include clearing for cultivation "the great rain-forest areas in South-East Asia, in the Congo and Brazil," irrigating drylands, and making the seas farms for fish.[88]

Bernal regarded postwar American efforts to export the techniques of birth control to the Third World with deep suspicion. Their professed rationale, to prevent population from outrunning resources, he thought transparently false, for "there has never been a time in the history of mankind when the doctrine of diminishing returns and exhaustible resources was more palpably nonsense." Existing resources of energy and technology, to say nothing of future scientific advances, could easily support the growing populations of the less-developed countries.[89] Humankind now possessed the means of "planned abundance," two words that, for Bernal, could not be separated.[90] A capitalistic class society, or a world dominated by such societies, would deny abundance to the majority of its people as communist planning would provide it.

Bernal admired the Soviet Union alike for its professed social ideals and for its environmental ones, as a progressive, classless, and egalitarian society and as one devoted to the useful application of science. He visited the USSR twice in the 1930s and several times again in the 1940s and 1950s, once to receive a Stalin Peace Prize, and he wrote an admiring obituary on Stalin's death in 1953, lauding the late dictator both as a statesman and as a thinker.[91] In 1939, he singled out the USSR as a model to the rest of the world as "exemplary" in its understanding of science as first and foremost a means of mastering the material world.[92] In 1952, he collaborated with several other British Marxists on a pamphlet titled *Man Conquers Nature: The New Soviet Construction Schemes.* He contributed a chapter on "The Engineer and Nature," focusing on the massive projects underway on the USSR's rivers for the purposes of

---

[88] Bernal, *The Freedom of Necessity*, 11–12, 417 (quotation); *World Without War*, 49, 71 (quotations), 72, 79.

[89] J.D. Bernal, "Science and Human Wants," *Science and Society* 20, #2 (1956), 107.

[90] Bernal, *The Freedom of Necessity*, 74–76.

[91] Brown, *J. D. Bernal.*

[92] Bernal, *The Social Function of Science*, 379–380.

irrigation, navigation, power generation, and flood control. "They have passed beyond the scale of merely modifying existing drainage systems here and there," Bernal noted, "and are approaching the time when they will design them over a whole area of a continent, to suit human needs. Already the rivers of northern Russia are being made to flow back over their water-sheds to feed the Volga and the dry lands to the south."[93]

Bernal dealt evasively with the phenomenon of Lysenkoist biology in the Soviet Union, perhaps finding its Promethean implications appealing enough to outweigh its flimsiness as science and the brutal methods by which it was imposed on the research community.[94] In the 1960s, he was still holding up the USSR to the Western countries as an environmental model. The exploitation of natural resources and the use of the earth's surface by capitalist societies had been wasteful and destructive, and their successes in remaking nature spotty and haphazard. "In the part of the world now saved from the operation of the free market and the monopoly trust the picture is very different." The USSR had undertaken "something radically new in the history of our planet: a deliberate effort to remake Nature and change geography in the service of mankind," irrigating or afforesting deserts, harnessing rivers, improving soils, cultivating "mountain wastes" and "making full use of the resources of the sea." The Soviet example marked a turning point in human history. "Geography can no longer be taken for granted, the world surface will henceforth be what man chooses to make it." China and other less-developed countries were following the Soviet example, and Bernal exhorted the USA to fall into line and compete with the Russians "at their own game—of changing Nature not for profit but for use."[95]

Bernal's hopes for the conquest of the earth, however grandiose, were modest compared with his broader visions, ones he stated quite early in his scientific career in a short book, *The World, the Flesh & the Devil* (1929) that he reprinted unchanged late in his life. The earth was itself only a part of an immensely larger cosmos, and the creation through science

[93] J.D. Bernal, "The Engineer and Nature," in *Man Conquers Nature: The New Soviet Construction Schemes* (London: Society for Cultural Relations with the USSR), 25–30 (quotation from 26–27).

[94] Brown, *J. D. Bernal*, 300–313, 364, 384, 407.

[95] J.D. Bernal, *Science in History*, revised edition (Cambridge, MA: MIT Press, 1969), 829–831, 964–969, 973–977, 1212, 1305–1306 (quotations from 965, 966, 968, 969).

and technology of "a world incomparably more efficient and richer than the present, capable of supporting a much larger population" was only an overture to a larger conquest that alone could secure the future of the human race. Bernal proposed that by some means—he suggested "space globes," miniature and modular worlds containing a group of human beings and all of the means needed for their sustenance—humankind should first colonize space and then set about reshaping and improving it in the very way it would already have done with its native planet. "Man will not ultimately be content to be parasitic on the stars but will invade them and organize them for his own purposes," and by making them vastly more efficient engines, would prolong the life of the universe as a home for intelligent life "to many millions of millions of times what it would be without organization."[96]

Bernal's like-minded radical contemporaries included the geneticist J.B.S. Haldane (1892–1964), who reserved most of his Promethean enthusiasm for the reshaping of humankind's biological nature, but occasionally turned his attention to its natural surroundings as well. He looked forward to the day when the land surface would be freed from the demands of agriculture by the development of manufactured food and instead "covered with rows of electric windmills" as the chief source of human power. Genetic engineering would produce nitrogen-fixing algae, vastly enlarging the meager natural fish supply of the oceans, and sand-fixing lichens would make possible the conquest of the world's deserts. But such a change as the deliberate farming of fish, he was happy to point out, would only be possible if the capitalist plan of "cut-throat competition" were replaced by national or international ownership and management of the seas, and if "scientific planning goes with scientific fertilization."[97]

The popularization of science, a frequent avocation of Haldane's, was another influential medium for the spread of Promethean ideas about the environment. Bernal in the 1930s favorably contrasted Soviet popular

[96] J.D. Bernal, *The World, the Flesh & the Devil: An Enquiry into the Future of the Three Enemies of the Rational Soul*, 2nd edition (Bloomington, IN: Indiana University Press, 1969), 14, 28.

[97] J.B.S. Haldane, *Daedalus; or, Science and the Future* (New York, NY: E. P. Dutton & Co., 1924), 23–27 (quotation from 24), 37–40, 59–63; *Science Advances* (New York, NY: Macmillan, 1948), 195–197.

science writing, with its emphasis on "how men can use science to struggle with nature and improve their condition," with the ineffective, genteel variety prevalent in Britain that sought only "to cause the reader to meditate on the mysteries of the universe." He singled out for praise the Soviet writer M. Il'in (1896–1953), whose work emphasized the need for a socialist organization of society to master the natural world.[98] Il'in drew the same contrast between a blind, uncontrolled nature and one regulated by intelligent human action that had played so large a role in Lester Frank Ward's thought. Nature, he wrote, "is self-willed and capricious. It knows neither purpose nor plan, does not consider what is good and what is bad for humankind." A nature "which humankind recreates according to a rational plan and with a rational purpose," "permeated by human thought," was a vast improvement. Both kinds were better than a third, mindless nature further disorganized and disrupted by the kind of unwise or irresponsible human interference Il'in saw exemplified in the capitalist USA. Such interference made the material world an all the more accurate mirror of the social one that had disrupted it. For unmanaged nature and capitalist society, Il'in wrote, again in Ward's vein, were alike in being realms of chaotic, uncontrolled forces and shot through with wastefulness, injustice, and destruction. Socialism meant a planned society that would also plan and rationalize the environment.[99] "According to a single scientific plan, for the benefit of all of human society, socialism governs the flow of rivers, creates artificial lakes and reservoirs, unites seas by canals, waters deserts, dries up wetlands, shifts the limits of cultivation northwards. There are no elemental forces that planned socialist labor cannot subjugate." Il'in speculated on the advances in science and technology that might put regional and global climates, long "uncontrolled by human beings," under intelligent direction for the first time. He drew pictures of the contrasting fates that such methods would encounter under capitalist and socialist societies. In the former, controlled by private interests in the quest for profit, they would only further disorganize

[98] Bernal, *The Social Function of Science*, 229, 237n22.

[99] M. Il'in (pen name of Il'ia Iakovlevich Marshak), *Pokorenie prirody* (orig. 1950), in *Izbrannye proizvedeniia v trekh tomakh*, vol. 1 (Moscow: Gosudarstvennoe izdatel'stvo khudozhenstvennoi literatury, 1962), 506–507 (quotations), 508–510.

the weather and worsen human life. In the latter, they would be directed by the government and used only for the general good.[100]

In the USA, the scientific popularizer Edwin E. Slosson (1865–1929), a friend and admirer of John Dewey and of the "New Historian" James Harvey Robinson and a longtime staff writer for the New York liberal weekly *The Independent*, linked the mastery of nature through the advances of science to progressive social ideals.[101] Conversely, he attacked the idolization of nature as "a reactionary spirit, antagonistic to progress and destructive to civilization." In a polemic titled "Back to Nature? Never! Forward to the Machine," he derisively paraphrased this outlook: "Praise the country and demean the city. Admire cliffs but make fun of skyscrapers. Extol forests and despise laboratories." "Chaos is the 'natural' state of the universe," he argued, and it was "only by means of applied science that the world can be made habitable and a decent human life made possible." Through it, man would by gradual steps "substitute for the natural world an artificial world, molded nearer to his heart's desire." Particularly impressed by the achievements of chemistry, Slosson approvingly cited Berthelot on its ever-increasing ability to create entirely new substances.[102] He devoted a book to the topic—describing its theme as that of "the conquest of nature, not the imitation of nature"—beginning with the synthesis of nitrogen compounds for use as fertilizers.[103]

Slosson expressed a high regard for a writer who gave more articulate expression both to the ideals of science and to those of social and environmental planning than perhaps anyone who has ever lived, placing him among the "Six Major Prophets" he studied in a book published

---

[100] M. Il'in, *Chelovek i stikhiia* (orig. 1947), in ibid., vol. 3, 20, 301 (quotations), 302–309. The prize-winning Soviet writer Nikolai Mikhailov similarly glorified the transformation of the land surface of the USSR by socialist planning and technology: Evgeny Dobrenko, "The Art of Social Navigation: The Cultural Topography of the Stalin Era," in Evgeny Dobrenko and Eric Naiman, eds., *The Landscape of Stalinism: The Art and Ideology of Soviet Space* (Seattle, WA: University of Washington Press, 2003), 189–199.

[101] On Slosson's career and his political views, see Ronald C. Tobey, *The American Ideology of National Science, 1919–1930* (Pittsburgh, PA: University of Pittsburgh, 1971).

[102] Edwin E. Slosson, "Back to Nature? Never! Forward to the Machine!," *The Independent* 101, #3703 (3 January 1920), 6, 38, 40.

[103] Edwin E. Slosson, *Sermons of a Chemist* (New York: Harcourt, Brace and Company, 1925), 116n (quotation); *Creative Chemistry: Descriptive of Recent Achievements in the Chemical Industries* (Garden City, NY: Garden City Publishing Co, Inc., 1919).

in 1917.[104] H.G. Wells (1866–1946) was the first distinguished English man of letters whose education had been a scientific rather than the traditionally classical and humanistic one. He was never a professional researcher in science, and only briefly as a young man a classroom teacher of it. Throughout his adult life he lived by writing books and journalism, winning fame for his fiction, his futuristic speculations, and his historical syntheses. Yet he drew constantly on his training and his amateur scientific pursuits for the matter of much of his fiction and his nonfiction alike, imaginatively as well as logically expounding many of the themes common to the scientific Prometheans.

In his 1934 autobiography, Wells succinctly stated the purpose that had guided him: "to rescue human society from the net of tradition in which it is entangled and to reconstruct it along planetary lines."[105] Even if one allows for an element of retrospective tidying-up, the summary fits the facts of his life well. It explicitly rejects a conservatism that holds to tradition as a safe and helpful guide; for Wells, tradition was usually a trap from which humankind needed to be extricated for its own good. It prescribes reconstruction, the word implying a conscious and planned effort rather than the outcome of independent actions coordinated, if at all, only by some set of forces beyond anyone's will. It prescribes action at the global scale, in accord with a diagnosis of troubles that are likewise global. And in the juxtaposed words "planetary" and "reconstruction" there is a hint that elsewhere in Wells is made as explicit as anyone could wish: that the planet is not merely where the reconstruction will take place, but one of the things to be reconstructed. Later in the same work, he defined his field as "human ecology" and his goal as "efficient world planning."[106] His work offers all the proof one might need that thinking of humankind as a species, its surroundings as its ecology, and its world as a planet is as compatible with Prometheanism as with environmentalism.

If conservatism entails a favorable attitude toward tradition and stability, it would be difficult to name a writer or thinker of equal stature

[104] Edwin E. Slosson, *Six Major Prophets* (Boston, MA: Little, Brown, and Company, 1917).

[105] H.G. Wells, *Experiment in Autobiography: Discoveries and Conclusions of a Very Ordinary Brain* (New York, NY: Macmillan, 1934), 549.

[106] Ibid., 552.

less conservative than Wells. His work is pervaded by a skeptical irreverence, owing much to his scientific training, towards the beliefs and practices that human beings have inherited from the past. He disbelieved as profoundly in the efficacy of a hidden hand keeping human affairs in harmony and equilibrium, or in the spontaneously evolved and selected social orders that Hayek would celebrate, as he did in any such agency of providence in nature. Left to themselves, neither society nor nature, he was convinced, produced the best of all possible worlds or anything like it. Rather, each blundered along towards no goal more satisfactory than a jerry-built mess. One of Wells's strongest emotions was a hatred of what an alter ego in one of his novels called "the folly and muddle that come from headlong, aimless, and haphazard methods," coupled with a passionate admiration for what is clean, sensible, and orderly and an equally passionate desire to transform the one state into the other by the application of mind. "I like order in the place of vermin, I prefer a garden to a swamp," he avowed in 1936.[107] He was equally disinclined in nature and in society to take what existed as a norm.

Nor did he believe that what existed could be long maintained even if that were desirable. From his scientific training, Wells had acquired a sense of the earth and the universe as subject to change without end, amid which human beings must make every effort to adapt in order to survive. By situating the present day, as he so often did, in the frame of the vast stretches of time that had preceded and that would follow it, he meant to remind readers of the novelty and transience of the world they took for granted and the impossibility of keeping it as it was: the supposedly eternal features of the planet no less than the ways of human life. "All the peace and fixity that man has ever known or ever will know," Wells wrote in a novel published in 1923, at a time when the line between fiction and persuasion was all but ceasing to exist for him, "is but the smoothness of the face of a torrent that flies along with incredible speed from cataract to cataract. Time was when men could talk of everlasting hills. Today a schoolboy knows that they dissolve under the frost and wind and rain

[107] H.G. Wells, *The New Machiavelli* (New York, NY: Duffield & Company, 1910), 42; lecture of 1936, printed in H.G. Wells, *The World Brain* (Garden City, NY: Doubleday, Doran & Co., 1938), 24.

and pour seaward, day by day and hour by hour. Time was when men could speak of Terra Firma and feel the earth fixed, adamantine beneath their feet. Now they know that it whirls through space eddying about a spinning, blindly driven sun amidst a sheeplike drift of stars."[108]

If nature could not be venerated for its fixity, neither could it be revered for its morals. Taking two courses with Huxley at the Normal School of Science seems to have been the formative intellectual experience of Wells's life.[109] He adopted Huxley's view of nature as fundamentally amoral—something one could hardly fail to recognize, both would have said, if one studied it with any degree of attention—and as offering no useful lessons in behavior to human beings. Morality, for both Huxley and Wells, was a human and cultural invention. Evidence of the viciousness rampant in nature cast a disconcerting light, which Wells was happy to focus, on its supposed origins in a benevolent creator. Far from seeking to palliate or gloss over the reality, he drove home the point by describing in detail the sheer nastiness of life among animals, plants, and microbes.[110] It was idle or worse, he implied, to ask human beings to care indiscriminately for a nature that cared nothing for them, for any other species, or for the individual members of any species. The ethical codes developed by human beings were real and right for them. Those that Wells advocated involved egalitarianism of opportunity and of obligation and the right of all to "food, shelter, and leisure" guaranteed by a planned economy and by the collective, not-for-profit ownership and management of the earth's resources.[111]

And if people could invent morals that improved on nature's, they could improve on it in other ways as well. They could, for instance, make

[108] H.G. Wells, *Men Like Gods* (New York, NY: Macmillan, 1923), 255.

[109] Wells, *Experiment in Autobiography*, 159–163; Michael R. Page, *The Literary Imagination from Erasmus Darwin to H. G. Wells: Science, Evolution, and Ecology* (Burlington, VT: Ashgate, 2012), 151–154; H.G. Wells, "Human Evolution: An Artificial Process" (1896), reprinted in *H. G. Wells: Early Writings in Science and Science Fiction*, ed. Robert M. Philmus and David Y. Hughes (Berkeley, CA: University of California Press, 1975), 211–219; H.G. Wells, "Morals and Civilization" (1897), reprinted in ibid., 220–228.

[110] W. Warren Wagar, *H. G. Wells and the World State* (New Haven, CT: Yale University Press, 1961), 65–71; for some particularly vivid examples, see H.G. Wells, *The Undying Fire: A Contemporary Novel* (New York, NY: Macmillan, 1919), 81–95, 187–188, 192.

[111] H.G. Wells, *The Open Conspiracy: Blueprints for a World Revolution* (Garden City, NY: Doubleday, Doran and Company, 1928), 34.

it vastly more productive. Wells was at one with Ward in his sense of nature's inefficiency. He expected that "an abundance out of all comparison greater than the existing supply of things" would ensue when scientific management had taken "the infinite wastefulness" of nature in hand and corrected it. "It's only when you come to artificial things," says an admirable character in one of his novels, who is juxtaposed against an aimless and nature-loving esthete, "such as a ploughed field, for example, that you get space and health and every blade doing its best." Like Ward, Wells equated economic competition with that of nature—in the tropical forest, "what is alive is either murdering or being murdered"—in both its waste and its cruelty.[112] Reclus's Prometheanism had been restrained by a deep admiration for many of the beauties of nature, capable of human embellishment though he thought them. Wells professed much less attachment even to the most widely admired of scenic spots. He wrote a provocative account of his impressions on a visit to Niagara Falls in 1906. He mentioned a frequent complaint by visitors of that era: that the natural beauties of the falls had been "destroyed beyond recovery by the hotels, the factories, the power houses, the bridges and tramways and hoardings that arose about it." They, indeed, made a sordid and hideous sight, he granted. But he doubted that, even in its pristine state, the cataract had offered anything qualitatively different from what many other, better-preserved waterfalls did. "The real interest of Niagara for me was not in its waterfall, but in the human accumulations around it," some of them inspiring rather than repulsive, and the most impressive thing he saw there was indoors in an entirely man-made setting: "[t]he dynamos and galleries of the Niagara Falls Power Company," clean and its "softly-humming turbines" clean, beautifully made, and "starkly powerful." They spoke persuasively to him of "the loss of all the accidental unmeaning beauty" of nature "that is going for the sake of the beauty of fine order and intention that will come." It was a minority view, he acknowledged. In most of the discussion then going on about the preservation of Niagara, it seemed to be taken for granted "that a voluminous waterfall is necessarily a thing of incredible beauty, and a human use is necessarily a

[112] Ibid., 45–46; H.G. Wells, *Joan and Peter: The Story of an Education* (New York, NY: Macmillan, 1918), 23.

degrading use." He begged to differ, and he felt confident that the future would agree that keeping the Falls untouched for their scenic sake would be "a Titanic imbecility of wasted gifts."[113]

For society needed all the resources, organizational and physical, that it could find. "There is no reason whatever," Wells wrote, "to believe that the order of nature has any greater bias in favor of man than it had in favor of the icthyosaur or the pterodactyl." Nature's own blind, uncontrolled workings posed a variety of dangers to the survival of the human race. Wells's story "The Raiders" offered a nightmare of humankind suddenly beset by the natural adversaries that evolution is more than happy to raise up against it, as packs of large tentacled sea creatures that have acquired a taste for human flesh begin to attack bathers along the coast of England. "The Empire of the Ants" begins in a situation of apparent farce. The commander of a Brazilian gunboat has been ordered, he suspects as a form of mockery by his superiors, to steam up one of the remote tributaries of the Amazon to assist villagers who are being killed or driven from their homes by ants. But, when encountered, the ants turn out to be like nothing the captain or crew have encountered before. Unusually large and capable of overwhelming and stinging humans to death, they are, still more frighteningly, deliberate, systematic, and intelligent in their actions. The gunboat retreats after cannonading the shore in a sorry and futile show of dignity, and the story ends with a report of the ants' current depredations, as best they can be made out. They have occupied sixty miles of riverbank, killed or driven off all of the human occupants, "boarded and captured at least one ship," and begun to show remarkable skills in the use of fire, metals, and bridge and tunnel construction that makes them a danger, the narrator thinks, not just to tropical South America but before long to Europe. With the chance appearance of intelligence among their other qualities, added to the efficiency of their social organization, they are "new competitors for the sovereignty of the globe" that human beings thoughtlessly took to be their assured birthright.[114]

[113] H.G. Wells, *The Future in America: A Search After Realities* (London: Chapman & Hall, 1906), 71–78.

[114] H.G. Wells, *The Fate of Man* (New York, NY: Longmans, Green & Co., 1939), 247; "The Raiders," in *The Short Stories of H. G. Wells* (Garden City, NY: Doubleday, Doran & Company, 1928), 410–420; "The Empire of the Ants," in ibid., 88–104 (quotations from 103, 104).

And perhaps Wells's best-known work, *The War of the Worlds*, centered on a similar threat, conquest by another form of life superior in intelligence and organization, this time originating from elsewhere in space. The inhabitants of Mars, "intellects vast and cool and unsympathetic, regarded this earth with envious eyes" from their own steadily colder and more drought-stricken world. They saw "our own warmer planet, green with vegetation and grey with water, and a cloudy atmosphere eloquent of fertility, with glimpses through its drifting cloud wisps of broad stretches of populous country and narrow, navy-crowded seas." What wonder that, being able to do so, they sent an expedition to begin its subjugation? Wells anticipates and answered a likely objection. Would a more evolved, highly developed race be so malevolent or insensitive as to brush aside the earth's human population? But to the Martians, Wells observed, not only was the acquisition of a new world a matter of sheer survival, but their attitude toward a species of such inferior abilities to their own could easily be understood. One needed only to remember how the most technologically advanced portions of the human race had dealt even during the nineteenth century with the lower animals or with the weaker human societies that had stood in their way. The Martians were successful Prometheans, until felled by the unexpected pathogens they encountered on earth; human beings, at least to date, were far less successful ones and needed, for their own safety, to catch up.[115]

Yet another ever-present threat to human survival lay in the possibility of a collision with some wandering cosmic body—a comet, a star—or even a near-collision sufficiently close to wreak havoc through the gravitational pull of the passing world. The story "The Star" describes a global cataclysm brought about by the passage close to the earth of a wandering celestial body. *In the Days of the Comet* (1906) ends more happily, but only because the comet's tail that envelops the earth contains a gas that enriches the atmosphere and enlightens humanity, the undeserved beneficiary of blind natural luck.[116] Nothing, to Wells's mind, could give humankind a chance for survival in such a universe but the use of its

[115] H.G. Wells, *The War of the Worlds* (London: Heinemann, 1898), 2, 4–5.

[116] H.G. Wells, "The Star," in *The Short Stories of H. G. Wells*, 631–642; *In the Days of the Comet* (London: Macmillan, 1906).

one real strength, its intelligence, to develop the technological capacity, first to deal with any challenges confronting it on earth, and eventually to escape the earth's limits and plant human life on other worlds: each a social as well as a technological challenge.

If *The War of the Worlds* is not Wells's best-known work, that distinction could only belong instead to his first successful novel, *The Time Machine* (1895). Though it can be read either way, it makes more sense when it is considered as an urgent dystopian warning than as a forecast of what must inevitably come to pass, just as the tale of the Martian invasion is best considered an admonition to humankind to strengthen its technological muscles and prepare for the possible worst. *The Time Machine* extrapolated the long-term consequences of the class division of nineteenth-century England, first to a superficially charming future that hides a nightmarish reality underneath, and then many ages further to an earth in which nature has eradicated humankind altogether. In it, Wells also hinted that the conquest of terrestrial nature, the softening of the climate and the cornucopian bounty of resources, will lead to human degeneration by taking away the spur of necessity: an anti-Promethean note that vanished from his subsequent work.[117]

Though, as often happens, Wells's dystopias have proven more memorable—or at any rate are better remembered—than his utopias, he traded in both commodities. Menacing the reader on the one hand with nightmares of what might come to pass if his diagnoses of the human condition were ignored, with the other hand he proffered tales of what might be possible if they were taken seriously and acted upon. His utopias have a distinctly Promethean cast, the later ones more stridently than the earlier. The earliest of them, *A Modern Utopia* (1905), proposed "society as an organization for the conversion of all the available energy in nature to the material ends of mankind" and judged the earth's resources, if society were organized properly, fully capable of meeting all of humankind's needs.[118] In 1908, Wells presented Mars as an object lesson for the earth. Writing of the beings whom he, in line with Lowell's theories, supposed to occupy it, Wells described them as "creatures of sufficient energy and

[117] H.G. Wells, *The Time Machine* (London: William Heinemann, 1895).

[118] H.G. Wells, *A Modern Utopia* (London: Chapman & Hall, 1905), 84.

engineering science to make canals beside which our greatest human achievements pale into insignificance." They "have taken Mars in hand to rule and order and cultivate systematically and completely, as I believe some day man will take this earth."[119]

A quirk of space-time catapulted a small set of human beings from the present day into the world of *Men Like Gods* (1923), which the visitors found occupied by a more enlightened humanity than their own. Not only the peace, plenty, and social harmony characteristic of utopia, but an intelligent Prometheanism had carefully preserved what was good in nature and done away with the rest. The rest had included such "tiresome and mischievous species" as parasites of all kinds, rats, wasps, flies, mosquitoes, "weeds and vermin and hostile beasts … driven out of life by campaigns involving an immense effort and extending over many generations." Species, useless in themselves, that helped sustain useful ones were retained, however, and none was exterminated without a searching inquiry into its possible value and without a small reserve being maintained for its possible reintroduction. The useful species had been made much more useful by a systematic program of genetic engineering that created new plants of great productivity or great beauty and cured the formerly carnivorous animals of their unseemly tastes. Forests that on earth were crippled and misshapen by competition for survival and by the attacks of pests were managed so that they grew to their full healthy potential. Gigantic dams and reservoirs with which the Utopians had graced the uplands furnished an abundance of power—"every torrent, every cataract was working a turbine"—and, like all of their other works, embellished rather than defaced the landscape with the superior effects of human art and "a bolder, more delightful design" than nature had ever produced. One of the sympathetic newcomers sums up their efforts: the Utopians had "tamed the forces of nature and subjugated them altogether to one sole end, to the material comfort of the race."[120]

Wells's most ambitious utopia was his 1933 future history of the conquest of darkness and chaos by a rational, scientific world state, *The Shape*

[119] H.G. Wells, "The Things That Live on Mars," *Cosmopolitan Magazine* 44, #4 (1908), 335–342 (quotations from 335, 340).

[120] Wells, *Men Like Gods*, 92–94, 98, 178, 261–262.

of Things to Come. In the year 2060, scientists can set about as never before to correct the "fundamental poverty of terrestrial existence" that the unobservant had long thought of instead as "the alleged bounty of nature." They can envision the creation of "such a plenty and wealth of life on our planet as the whole universe had never dreamed of before this time." They can design new plant species for use and decoration and undertake the biological reengineering of insects and larger animals "until they all come into a tolerable friendship with ourselves." The control of the weather is still beyond reach, though its accurate forecasting is not, but the land surface is a vast worksite. "An immense series of enterprises to change the soil, lay-out, vegetation and fauna, first of this region and then of that, will necessitate a complete rearrangement of the mines, deep quarries, road network and heavy sea transport of the globe." These more pressing necessities taken care of, the more distant future is expected to undertake "geogonic planning" of a far more profound character "to alter the terrestrial contours" and "remodel the world," redrawing the map of land and sea and the distribution of mountains and lowlands. Wells's repeated expressions of concern about overpopulation had less to do with any belief in the inexorability of natural limits than with his conviction that human numbers, unchecked, threatened to outgrow the meager capacity of a wasteful, ineffective, unplanned economic system of production to provide them with sustenance. The rationalization of society and resource use in *The Shape of Things to Come* eventually permits a substantial increase in human numbers as well as in lifespan.[121]

Wells shared much intellectual ground with the great early English Promethean Francis Bacon. Both saw the ultimate goal of science as usefulness in improving human life. Late in his life, Wells defined "effective knowledge" as the goal he had always been seeking and been urging the world to acquire.[122] There was much knowledge that was not effective, as

---

[121] H.G. Wells, *The Shape of Things to Come* (orig. 1933; New York, NY: Macmillan, 1945), 388–389, 395, 396, 397. For the Promethean character of Wells's nonfiction science writing from the same decade, see Peder Anker, *Imperial Ecology: Environmental Order in the British Empire, 1895–1945* (Cambridge, MA: Harvard University Press, 2001), 110–114.

[122] Richard Nate, "Scientific Utopianism in Francis Bacon and H. G. Wells: From Salomon's House to The Open Conspiracy," in Barbara Goodwin, ed., *The Philosophy of Utopia* (London: Frank Cass, 2001), 172–188; Wells, *The Fate of Man*, 1.

Bacon had emphasized in criticizing the prevalent learning of his time. It took three forms, he had written, all of them objectionable in their own ways: contentious, delicate, and fantastic learning.[123] Fantastic learning was magical lore, science without scientific grounding, and its twentieth-century cognates interested Wells little. Delicate learning was humanistic trifling, useless erudition for its own sake, grace and embroidery of style hiding a poverty or an absence of purpose. Its twentieth-century cognates interested Wells very much indeed and engaged him emotionally as well, his dominant emotion being a raw hostility toward their irresponsible uselessness in the face of the great human tasks.[124] A national program of education centered on the delicate learning of the ancient literary classics, like that of England in his time, raised the noxiousness of the individual offender into a system, according at best a secondary place to the study of the natural world with which humankind must struggle to prosper and of the social world whose adequate or inadequate design would determine the outcome of the struggle with nature. Baconian contentious learning Wells saw embodied in the absurdities of the English legal system, on which the suffocating hold of tradition imposed an exasperating conservatism of both doctrines and procedures.[125] In his novel *The World Set Free* (1914), he juxtaposed English law with science (not English science; science, to his mind, and to its great credit, had no nationality; the one parochialism he tolerated and indeed promoted was loyalty to the human species and to the cause of its survival) to point up the sorry worthlessness of the former, whose anachronistic forms and costumes perfectly symbolized the backwardness of its doctrines and procedures. As opposed to such varieties of ineffectiveness, effective knowledge was what put power in human hands and ensured that it would be used for human betterment. *The World Set Free* opened by illustrating what the first without the second would bring. Abundant energy from the atom that ought to have been an unmixed blessing first calamitously disrupted the world's

[123] Francis Bacon, *The Philosophical Works of Francis Bacon*, ed. James Spedding, vol. 3 (London: Longmans & Co., 1861), 282.

[124] Wagar, *H. G. Wells and the World State*, 239–240, 239n.

[125] For Wells's attitude towards the British legal system and the "legal mind" more generally, see H.G. Wells, *The Discovery of the Future* (London: T. Fisher Unwin, 1902), 7–11; Wells, *Experiment in Autobiography*, 553.

economy and then led to global warfare of unprecedented destructiveness, because of the utter inadequacies of the economic and social systems to turn it to human benefit. The disaster occurred because the world was "not really governed at all." In the new and better world that finally emerged after much chaos: "Contentious professions … man the warrior, man the lawyer, and all the bickering aspects of life" have disappeared.[126]

What was to be done? Plainly not a continuation of business as usual, nor an earnest consultation of the wisdom of the past. Seeing, as he did, a pervasive inefficiency perpetuated by routine as characteristic of most of society, Wells was perhaps as far from Edmund Burke and his distrust of innovation as any major writer has been. In a book published in 1914, he defined his world view by contrast with two others. One was the mindset of the "Conservators" (a word he preferred to "Conservatives," lest he be supposed to be referring narrowly to the members of the British political party alone). They thought the ways that had been characteristic of human experience to date "the only proper and decent life for the great mass of humanity. … Their attention to the forces of change is necessarily a hostile attention," and the hostility extended to science, especially organized science, which they feared as a prime source of change. Conservators in this sense might include avowed socialists. They included, in Wells's estimation, William Morris, whom he thought a "profoundly reactionary" thinker, as sworn an enemy of fundamental novelty in human ways of life as many who labeled themselves conservatives. Wells's Progressives, those who thought a different and better future life possible, covered the range between two polar types. "Planless Progressives" (whose prophet Hayek would become) were individualists who disliked organization. They trusted in hidden harmonies or a process of selection to foster desirable changes and suppress harmful ones. Wells thought them naïve, and his own sympathies lay far towards the other end of the scale, among those he called "Constructors," "those types which believe supremely in systematised purpose." In their view, the forces that were unavoidably disrupting the old ways so prized by the Conservators needed "to be controlled by a collective effort implying a collective design, deflected from

[126] H.G. Wells, *The World Set Free* (New York, NY: E. P. Dutton, 1914), 56–62 (quotation from 56), 248.

merely injurious consequences and organised for a new human welfare upon new lines." Change was not necessarily good, but the forces driving it had vast potential to produce good if they were mastered and directed. Liberal free-market capitalism, a planless progressivism, offered far too little mastery or direction. The book closed with a hymn to the sovereign power of intelligence that had allowed constructive humankind to free itself from and in turn dominate the nature that had blindly happened to produce it. In times to come, it would set out from the earth to conquer other planets and suns, "flying swiftly to unmeasured destinies through the starry stillness of space."[127]

The war that broke out in the same year, to say the least, did nothing to weaken Wells's convictions about the failure of human institutions to keep pace with the energies they were needed to regulate and about the urgency of a rational reordering of the world's affairs. As one student of his work, Warren Wagar, observed, Wells had the misfortune in his later years and after his death to be caricatured as a prophet of inevitable progress. What he wrote, in Wagar's words, is easily summarized and far different: "first, that Homo sapiens had made substantial progress in social organization and in controlling his physical environment since prehistoric times; second, that it might lie within his power to make still more progress along the same lines; and third, that nothing guaranteed man's ultimate success or failure. … If man's will failed, or if fate turned against him, he could be wiped out in a wink."[128] Two other readers of Wells have aptly defined as one of his central themes—it would perhaps not be an exaggeration to call it *the* central theme, or at least the most fundamental—"the precarious position of man in the universe."[129] Until the despairing last year or two of his life and his final book *Mind at the End of Its Tether* (1946), Wells was no believer in the inevitability either of triumph or of doom. Which was more likely (there were no certainties either way) depended immensely on the choices that would lead to one or the other. The route of Prometheus offered far better odds than that of

[127] H.G. Wells, *Social Forces in England and America* (New York, NY: Harper & Brothers, 1914), 125, 126, 130–131, 414–415.

[128] Wagar, *H. G. Wells and the World State*, 81–82.

[129] Robert M. Philmus and David Y. Hughes, "Precarious Man," in Philmus and Hughes, eds., *H. G. Wells: Early Writings in Science and Science Fiction*, 148.

complacent reliance on the providence and benignity of the universe, but how were Promethean energies to be released and applied? Wells's deepest argument for a world government, a cause he emphasized ever more strongly over the course of his life, was that the species could only hope to survive against the many challenges besetting it through a rational direction of all of its activities, energies, and resources. The invasion from Mars, his narrator reflected at the end of *The War of the Worlds*, had had its beneficial side. It had taken away the "serene confidence in the future" that made for complacency and laziness, and it "did much to promote the conception of the commonweal of mankind."[130]

For Wells, as for Ward, the world (and the earth) suffered from far too little government, certainly not from too much. He spoke in 1910 of "the state-making dream" of directed, coordinated effort toward "a world better ordered, happier, finer, securer" (and one whose achievements would include "great roads engineered marvellously, jungles cleared and deserts conquered"). By the same token, for Wells as for Ward it suffered from too much reliance on the competitive market: "surely nobody in his senses," Wells wrote, "believes that the supply and distribution of staple commodities about the earth by irresponsible persons and companies working entirely for monetary gain, is the best possible method from the point of view of the race as a whole."[131] In their senses or not, there were such believers, and one of them, Friedrich Hayek, living in England from 1931 to 1950, cast an equally disapproving eye on the "men of science," as he called them, including Bernal and Haldane and, as their most eloquent prophet, Wells. He regarded them as exemplary of a tendency among students of the natural world, one he traced back to its appearance among the engineers and technocrats of the Saint-Simonian school, to want to impose a program of rational planning and reconstruction on the world of human society, usually in something like the form of "the socialist ideal of a centrally planned economy." Scientism, though not necessarily science, Hayek thought, was always prone to the dream—"the fatal conceit," he called it, an illusion dangerous when not merely futile—of progress achieved by replacing the apparent disorder,

[130] Wells, *The War of the Worlds*, 300.

[131] Wells, *The New Machiavelli*, 4–5; *The Open Conspiracy*, 49–50.

inefficiencies, and illogicalities of all societies as they are with something more rationally designed and more just and morally attractive.[132] The dream, for such scientists as Ward, McGee, Bernal, and Wells, did not stop at the borders of society; it encompassed the world of nature that they studied as well. The twin projects of reforming society and reforming nature on more humane and efficient lines seemed to them to go quite comfortably together.

[132] Bruce Caldwell, *Hayek's Challenge: An Intellectual Biography of F. A. Hayek* (Chicago, IL: University of Chicago Press, 2004), 235–236; Hayek, *The Fatal Conceit*, 55; F. A. Hayek, "The Collectivism of the Scientistic Approach," in *Studies on the Abuse & Decline of Reason*, ed. Bruce Caldwell, *The Collected Works of F. A. Hayek*, vol. XIII (Chicago, IL: University of Chicago Press, 2010), 117–125.

4

The Prophetic Prometheans: Envisioning a New World and New Earth

One of the reasons scientists put too much faith in social planning, Hayek suggested, was a bias that their training had given them. They thought of social phenomena in the way they thought of natural ones, as the same reality for all observers, objectively describable, objectively valuable, and therefore easy for experts to manipulate for the better. But imagine, Hayek wrote—making use of an old literary device—having to explain the social world, the world of human-made institutions and ways of life, to newcomers from Mars. One makes it comprehensible to them only by invoking subjective interests and values, ones lying beyond the reach of the senses and of quantitative scientific measurement, and yet it would be no less rational for that.[1]

In 1920, the year of his death at the age of 83, the American writer William Dean Howells devoted one of the regular columns he wrote for *Harper's Monthly* to his experience in trying to explain the world to a pair of just such extraterrestrial visitors. Some recent news reports of flashing lights on Mars that might be signals from its inhabitants had reawakened

[1] Hayek, *Studies on the Abuse & Decline of Reason*, 123.

W.B. Meyer, *The Progressive Environmental Prometheans*,
DOI 10.1007/978-3-319-29263-2_4

interest in Percival Lowell's theories of a canal-building civilization on the red planet. So Howells was pleased when a two Martians showed up at his door in New York City to acquaint themselves with the ways of life on earth. Yet, when he tried to explain recent world events to his guests, he could find little of the underlying logic and sense that Hayek would have expected them to contain. The major powers of the globe had just devoted four years to a huge, bloody, and destructive war, for reasons that even the most sympathetic outsiders had to find difficult to grasp. Seeking a less awkward topic, Howells drew his visitors' attention instead to the cityscape of Manhattan. He hoped that its prodigies of construction might impress them as at least a faint shadow of "[t]he public works which you have carried through at home in the construction of the canals." Unfortunately, the new topic proved a difficult one for other reasons. The two Martians, it turned out, were outspoken socialists. Learning that they had been engaged to address a public meeting on their planet and its civilization, Howells cautiously advised them—for this was the time of the "Red Scare" and the Palmer raids in the USA—to confine themselves to describing the canals as works of physical engineering, and above all not to "touch upon moral or economical affairs." Both told him flatly that the canal system and the collectivist system of society that had fashioned it could not be discussed separately. When the time for the lecture arrived, they disregarded his well-meant warning, and their radical opinions caused an uproar. The authorities took them into custody, and, not knowing how to send them back to Mars, deported them to what seemed the next most suitable place: Russia, with its newly established Bolshevik regime.[2]

This brief late sketch echoed Howells's early and more extended use of utopian fiction both to criticize American institutions and to connect the beneficial engineering of the environment to a better organization of society. Conventional in his political and economic outlook as a young man, he began a leftward progress during the 1880s that would end in his espousal of a democratic Christian socialism.[3] The first of his utopian

[2] W.D. Howells, "Editor's Easy Chair," *Harper's Monthly Magazine* 140 (April, 1920), 710–712.

[3] See, e.g., Edwin H. Cady, *The Realist at War: The Mature Years, 1885–1920, of William Dean Howells* (Syracuse, NY: Syracuse University Press, 1958) and Robert L. Hough, *The Quiet Rebel: William Dean Howells as Social Commentator* (Lincoln, NE: University of Nebraska Press, 1960).

novels, *A Traveller from Altruria* (1894), used the words and deeds of a visitor from an imagined island nation of the southern hemisphere to highlight the inequalities of American society, its brutal social and economic competitiveness, and the hypocrisy of its professed ideals. A series of letters from the Altrurian visitor continued the commentary from this angle, and a third work, *The Eye of the Needle* (1907), took the visitor's newlywed American wife to Altruria itself to admire its progressive, democratic, cooperative, and egalitarian ways.

In its dealings with the environment no less than in its own organization, Altruria offered a model of intelligent reform. The redesigned landscape of Central Park made Howells' visitor think of the way in which nature in Altruria was made, on a much larger scale, to "yield herself to the enlightened will of man" and of the way "we are used to seeing the powerful machinery of our engineers change the face of the landscape," making "forests vanish in a night."[4] Altruria had been "long since cleared of all sorts of wild beasts," the countryside irrigated by the control of the runoff from the mountains, and the land cover "taken back into the hand of man." The Altrurians had even transformed their climate, cutting through a peninsula that had formerly blocked a warm current from approaching the coast. "Whole regions to the southward, which were nearest the pole and were sheeted with ice and snow, with the temperature and vegetation of Labrador, now have the climate of Italy," writes the Altrurian traveler's American-born wife, "and the mountains, which used to bear nothing but glaciers, are covered with olive orchards and plantations of the delicious coffee which they drink here."[5] Her husband found it a sad comment on their backward state of mind that "Americans have as yet no conception of publicly modifying the climate, as we do," but relied instead, inefficiently and inequitably, merely on the control of the climate indoors.[6] "The great outdoors is pruned, controlled, and

[4] W.D. Howells, *Letters from an Altrurian Traveller, 1893–1894* (Gainesville, FL: Scholars' Facsimiles and Reprints, 1961), 23.

[5] W.D. Howells, *Through the Eye of the Needle: A Romance; With an Introduction* (New York, NY: Harper & Brothers, 1907), 162–163, 164, 165, 170. The first Altrurian book described the same project of climate modification: W.D. Howells, *A Traveler from Altruria: Romance* (New York, NY: Harper & Brothers, 1894), 276–277.

[6] Howells, *Letters*, 88.

humanized" in Altruria, as one student of Howells's work has observed, under the direction of an active and enlightened state.[7]

A contemporary of the novelist's, Ignatius Donnelly (1831–1901), also turned his hand to utopian fiction during a career devoted mostly to other pursuits. After achieving note as a congressman, town promoter, and speculative historian, Donnelly helped in the early 1890s to found the Populist Party, with the later Progressives one of the two major left-wing movements in American politics between the Civil War and the Great Depression. He took a prominent part in the Populists' 1892 national convention and drafted the preamble to their platform.[8] In the same year, he published a utopian novel dealing with the currency question central to the election campaign. By discovering a way to manufacture gold, Donnelly's hero not only solved the pressing problem of deflation in the American economy—deploying the power of invention to make a critically important natural resource abundant—but after many vicissitudes was able to impose a just and democratic government on the entire world, whose fruits would include some Promethean triumphs in the effort to "vastly enlarge man's dominion over the earth": the clearing and cultivation of wilderness, the drainage of wetlands, and the conquest of deserts—even of the Sahara—by weather and climate modification.[9] As a member of the national House of Representatives for three terms in the 1860s, Donnelly had been a strong advocate of government aid for waterway improvements and railroad expansion.[10] He had also been Congress's most eloquent advocate of measures to transform the trans-Mississippi West still more profoundly. The curse of aridity, he declared in a speech in 1868, threatened to prevent the widespread settlement of

[7] Jean Pfaelzer, 1984. *The Utopian Novel in America, 1886–1896: The Politics of Form*, Pittsburgh, PA: University of Pittsburgh Press, 1984), 69.

[8] Martin Ridge, *Ignatius Donnelly: The Portrait of a Politician* (Chicago, IL: University of Chicago Press, 1962).

[9] Ignatius Donnelly, 1892. *The Golden Bottle; or, the Story of Ephraim Benezet of Kansas* (New York, NY: D. D. Merrill, 1892), 219–220, 224–225, 266, 267–268. Populism, long caricatured as anti-scientific and backward-looking, was closely akin to the later Progressive movement in its enthusiasm for technology: Charles Postel, *The Populist Vision* (New York, NY: Oxford University Press, 2007).

[10] Martin Ridge, "Ignatius Donnelly: Minnesota Congressman," *Minnesota History* 36, #5 (1959), 176–180.

the region, but it was one that human powers could dispel. Irrigation was one means for doing so; not only would its channels bring the crops the water that they needed, but evaporation from the network would moisten and permanently improve the climate. Moreover, Donnelly argued, the widespread planting of forests would likewise help to make the western climate more humid and more suited to agriculture. He called for joint public and private efforts "until all the torrents of the mountains and all the rivers of the plains are brought under the control of man, and nearly all the level land of these great wastes are [sic] rendered fertile and populous and prosperous," in conformity with the general principle that he enunciated: "The world was made for man, and he is its master."[11]

Woman, too, and perhaps even woman principally, Charlotte Perkins Gilman (1860–1935) might have added. An independent prophet of the American Progressive spirit, she energetically expounded its principal ideas in both fictional and non-fictional form. Like her contemporaries, the progressive conservationists, she particularly emphasized the inferiority of the natural to the artificial. "Nature's economy," Gilman wrote in her 1911 treatise *The Man-Made World*, "is not in the least 'economical.'"[12] She dedicated the book to Lester Frank Ward, whom she deeply admired: for his advocacy of women's equality, for urging the usefulness and benefits of state action in an era dominated by advocates of laissez-faire, and for championing the rationally designed over the natural. If anything, she intensified his moral case against nature and against an ungoverned human society analogous to it. Nature seemed to feel no distaste for parasites, for those who lived on the work of others, Gilman observed, even of the most horrible kinds. "She seems as fond of cancers as of lungs and livers, as willing to develop lice as nightingales," and laissez-faire in human society produced an exactly analogous menagerie of similar pests and victims, including "fat tax-collectors and lean peasants."[13]

---

[11] "Irrigation and Forest Culture: Speech of Hon. I. Donnelly, of Minnesota, in the House of Representatives, July 15, 1868," *Congressional Globe*, 40th Congress, 2nd session, Appendix, 474–478.

[12] Charlotte Perkins Gilman, *The Man-Made World; or, Our Androcentric Culture* (New York, NY: Charlton Company, 1911), 228.

[13] *The Forerunnner* (a magazine that Gilman almost entirely wrote herself between 1909 and 1916), 5 (1914), 201–202.

But the adoption of a socialist economy and an egalitarian polity, Gilman argued, could destroy the processes that created such pathological types. By the same token, the human race, of all species, was "incomparably the best fitted to change the environment to meet its own needs." "Humanity," Gilman wrote, "has improved upon nature beyond measure."[14] As evidence, she pointed again and again to "great irrigation works and ocean-joining canals," to the drainage of wetlands, to the immense improvement of the qualities of domesticated animals and plants by selective breeding, to the artificial raising of fish in rivers and along coastlines.[15] Human actions, she granted, had also, in many places, damaged or degraded the earth's surface, in wasteful and destructive deforestation, soil erosion, the silting up of harbors, severe water and air pollution, the expansion of deserts, and the foulness and disease of crowded cities.[16] But these proved nothing, to Gilman's mind, against the value of intelligent and purposeful reconstruction of whatever in nature human beings found ill-suited to their purposes. Because a dysfunctional society could irresponsibly alter the environment for the worse did not mean that a healthy one could not vastly improve it.

It could, for example, bring the earth's other species under better control. Gilman objected vehemently to unnecessary cruelty, to the killing of birds for their feathers and of fur-bearing animals for their pelts, to the mistreatment of beef cattle and other livestock, and likewise to zoos, which she called "beast prisons."[17] But she acknowledged no good in the existence of whole species if human beings found them harmful or merely useless or offensive. She lauded efforts to eradicate flies, mosquitoes, and gypsy moths and indeed insects in general. "Except for the bee and the silk-worm, we could spare them all," she wrote, for they were "our worst, our most numerous foes," and not merely harmful but disgusting. She found them almost all "hideous and noisome: the slimy slug, the fat greedy grub, the crawling caterpillar, the moth that corrupts, the

[14] *The Forerunner*, 7 (1916), 15, 76.

[15] E.g., *The Forerunner*, 2 (1911), 168; 6 (1915), 6, 199; 7 (1916), 333.

[16] E.g., *The Forerunner*, 2 (1911), 94, 272; 3 (1912), 23–24; 4 (1913), 27; 5 (1914), 303.

[17] *The Forerunner*, 2 (1911), 67, 128–130; 5 (1914), 94–95; 6 (1915), 215–216.

mosquito that inoculates us with fever, and the fly that wipes diseases from its feet upon our faces."[18]

Gilman was as hostile to conservatism as Wells or Ward, if not more so. She saw numberless inefficiencies, inequalities, and abuses in human folkways perpetuated by habit and indoctrination and protected from change or criticism by doleful warnings about the dangers of the unknown and the safety and value of tradition. She applied the lesson as consistently to the world of nature as to that of society. A parable she wrote for her magazine *The Forerunner* in 1913 recorded a debate between innovators and defenders of the environmental status quo in an imagined setting of human prehistory. What the innovators decried in the world of nature—"the wild wood with its fierce beasts, the desert with its parching sun, the swamps, breeding disease, the rivers that flood and drown, the fruit that is poison and the flowers that die"—the conservatives defended as necessary evils that humans simply had to accept. Fortunately, the innovators were not deterred: "Some beasts they tamed and others they slew, and the forest became a tended garden, yielding undying wealth. To the desert they brought water, the swamps they drained, the rivers they turned at will to wide, safe watercourses, the life-blood of the land. The best fruit was made better and planted far and wide, the poisonous fruit was destroyed root and branch, and the flowers were nurtured till they grew on every side and died not."[19] In another parable, Gilman contrasted two sets of farmers. The "Devout Farmers" took what came. In good weather, "they Thanked the Lord," and when drought came, they prayed for rain. Other farmers "prayed not, but dug, making little ditches that ran from the big canals that flowed from the great reservoirs behind the mighty dams they had builded." "And the Devout Farmers bowed their heads beneath the heavy hand of the Lord, and became poor, while the others laid their own heavy hands upon the hoe-handle, and prospered exceedingly."[20] Imagine the wastefulness of leaving Niagara Falls undeveloped for power production, Gilman wrote elsewhere, because tradition

[18] *The Forerunner*, 3 (1912), 6; 4 (1913), 18, 131; 6 (1915), 221.

[19] *The Forerunner*, 4 (1913), 326–327.

[20] *The Forerunner*, 3 (1912), 20–21.

happened to teach that waterfalls were sacred and that meddling with them would bring disaster.[21]

"A country netted with perfect roads, its forests, rivers and food-bearing resources all intelligently conserved and developed" was Gilman's vision of a reshaped earth, the utopia of the progressive conservationists and of Howells's Altrurians. She foresaw a "Federal Bureau of Improvements," for whose summer jobs of planting useful forests, stocking waters with fish, and building roads and dams and tunnels city residents seeking a change of work and scene for their vacation would eagerly compete.[22] And she made these kinds of activities central to her three utopian novels: *Moving the Mountain*, *Herland*, and *With Her in Ourland*. The first told the story of John Robertson, who returns to the USA after three decades of coma and amnesia in a Tibetan village, finding its politics and economy reformed and much of its environment as well. In *Herland*, three young men discover an advanced society, composed entirely of women, isolated from the rest of the world; in the sequel, they visit the USA with their Herlandian partners. All three novels, in the literary historian Michael Bryson's words, reflected "Gilman's abiding faith in the early-twentieth-century goals of progress and the control of nature" and describe "an extraordinarily high degree of human impact on the environment."[23] All of the improvements that they described arose not from a system of competitive laissez-faire, but from the planned and coordinated effort of a united society. It laced the land with roads and navigable waterways to facilitate transportation. It exterminated harmful insects and many other dangerous and useless animal species.[24] It kept much of the land in forest, but not in old growth, for it was enlightened enough to scorn

[21] *The Forerunner*, 3 (1912), 191–192.

[22] *The Forerunner*, 4 (1913), 250; 3 (1912), 296–297.

[23] Michael A. Bryson, *Visions of the Land: Science, Literature, and the American Environment from the Era of Exploration to the Age of Ecology* (Charlottesville, VA: University of Virginia Press, 2002), 57, 76. Another scholar has observed that Gilman (Ellen Richards as well) was environmentally a conservationist and not a preservationist: K.R. Egan, "Conservation and Cleanliness: Racial and Environmental Purity in Ellen Richards and Charlotte Perkins Gilman," *Women's Studies Quarterly*, 39 #3/4 (2011), 77–92.

[24] Minna Doskow, ed., *Charlotte Perkins Gilman's Utopian Novels: Moving the Mountain, Herland, and With Her in Herland* (Madison, NJ: Fairleigh Dickinson University Press, 1999), 87, 91–94, 103–104, 163–164, 188–193, 232–233.

untamed wilderness as wasteful and slovenly and to replace the native vegetation with a thick cover of useful, food-bearing trees. The "cool, spacious, flower-starred, fruitful forests of this time," representing in reality "a truck farm," not a forest, had supplanted "the tangled underbrush, with crooked, crowded, imperfect trees struggling for growth," that used to encumber the soil, the "woods, ragged and thick with dead boughs, fallen trunks and underbrush, not touched by any forester" of the less progressive past.[25] Gilman's utopia, like Howells' Altruria, is green in the literal but not in the metaphorical sense. Towards the end of *Moving the Mountain*, Robertson visits his elderly Uncle Jake on the latter's hill farm and finds him a malcontent, a voice of unintelligent conservatism, and a domestic tyrant to boot. Uncle Jake expresses his disgust with all "these new-fangled notions," "wimmin votin' now," blacks who want to be "treated like white folks," and suchlike absurdities, and no less with the reformed environment: "The mountains ain't what they used to be, John. They've got the trees all grafted up with new kinds of foolishness—nuts and fruit and one thing'n another"[26]

Prophecies of a new world and a new earth combined, with directions on how to reach them, make up a third category of nineteenth- and early twentieth-century progressive Promethean thought. Authors of utopian fictions since the genre was born have sought to prod or entice humankind forward by drawing pictures of the happiness that it could substitute for the misery of the present day. It could only do so, they have argued, if it abandoned the cramping fetters of the past and undertook the necessary reorganization of society. There have been conservative and even reactionary utopias too, whose happiness lies in a past to which we must return, but it is not by chance that the genre is dominantly progressive. Akin in some ways to the religious promise of paradise, of a perfect world offered on the condition that people submit themselves to the discipline of the faith, utopia differs in being a human achievement rather than a divine gift, one that can only be brought about by the conscious, intelligent, and organized efforts of the human race. Forward-looking fictions of progressive social change enjoyed a particular burst of popularity in the

---

[25] Ibid., 121, 143, 160 (quotations); 105, 110, 141, 143, 158–160, 205, 214–215.

[26] Ibid., 144, 146.

USA during the last dozen years of the nineteenth century, among whose largely obscure authors Howells, Donnelly, and Gilman are of particular note for the prominence they achieved in other fields. Most of the rest shared not only their social but their environmental progressivism, their enthusiasm for reforming nature along with the institutions of human life.[27] But utopias need not be presented as stories, and some of the most compelling have not been. The nineteenth-century reader also enjoyed a wide choice among plans in nonfictional form for achieving the best of all possible worlds on a perfected planet.

The Frenchman Charles Fourier (1772–1837) developed one of the most influential proposals for progress in society combined with the radical remaking of physical geography. Fourier's first book, *Théorie des quatre mouvements et les destinées générales*, appeared in 1808 but received little attention, as did a second, two-volume work elaborating much the same material in more detail, *Traité de l'association domestique-agricole* (1822). In the early 1830s, though, the enthusiastic support of a band of disciples—some of them drawn from the disintegrating Saint-Simonian movement—began to make Fourierism what it remained for some decades, the single most visible form of utopian socialism in the Western world, more even than its chief competitors, Saint-Simonianism and the doctrines of the Anglo-American industrialist and radical social theorist Robert Owen (1771–1858).[28]

The three blueprints for bliss were hardly identical. The Saint-Simonian utopia was for the most part a centralized one, governed directly by the leading experts and authorities for the common good. Fourier's, like

[27] Neil Harris, "Utopian Fiction and Its Discontents," in Richard L. Bushman et al., eds., *Uprooted Americans: Essays to Honor Oscar Handlin* (Boston, MA: Little, Brown, 1979), 209–237. The generalization applies, despite some superficially green elements, to the best-known of these works, Edward Bellamy's *Looking Backward*: Thomas Peyser, *Utopia & Cosmopolis: Globalization in the Era of American Literary Realism* (Durham, NC: Duke University Press, 1998), 87–89 and Brian Allen Drake, *Loving Nature, Fearing the State: Environmentalism and Antigovernment Politics before Reagan* (Seattle, WA: University of Washington Press, 2013), 139–140.

[28] The fullest biography of Fourier and account of the development of his ideas is Jonathan Beecher, *Charles Fourier: The Visionary and His World* (Berkeley, CA: University of California Press, 1986). Fourier's two principal works were *The Theory of the Four Movements* (orig. 1808), trans. Ian Patterson, eds. Gareth Stedman Jones and Ian Patterson (Cambridge: Cambridge University Press, 1996) (hereafter, Fourier, *Theory*); and the *Traité de l'association domestique-agricole*, 2 vols. (Paris: Bossange, 1822) (hereafter, Fourier, *Traité*).

Owen's, was based on the self-governing social unit, one that Fourier called a phalanstery, consisting of roughly 300–400 families or 1000–1500 persons. They agreed, however, in their disdain for tradition and in their emphasis on progress in a more egalitarian direction. All maintained that the institutions—political, economic, social—and the ways of thinking inherited from the past were badly suited to the needs of the present and the future and must be replaced with ones better adapted to the changes society was undergoing. All, accordingly, presented theories of progress in the form of stages, each one superseding and surpassing most of what had come before, which was irrevocably vanishing into the past and neither should nor could be revived.

Saint-Simon wrote of the disintegrating age of feudalism and theocracy and its imminent replacement by one of organized industry. Fourier offered an astonishingly detailed chart of thirty-two specific eras of world history covering the 80,000 years in which he supposed that the globe would support human life, a schema he declared applicable not only to the earth but to all of the inhabited planets of the universe. Parts of the earth currently had reached the state of "civilization," with vestiges of the previous stages—initial confusion, primitivism, savagery, patriarchate, and barbarism—still scattered about in various mixtures. All of them, civilization included, still lay on the wrong side of the single most important transition in the evolution of a world, one from social infancy to maturity or from "incoherence" to "combination," which would take place when society left civilization behind to cross the steppingstone of "guaranteeism" to full social harmony. The one variable in Fourier's otherwise rigidly timed series of stages lay in this transition between civilization and what followed, which would take place only when people awakened to the secret that would bring order and prosperity out of chaos. Then, and then alone, humankind as a whole would shift from being unhappy to being happy. It had already waited too long, Fourier asserted, through 2500 years of fruitless philosophic speculation, for the unlocking of the secret, which had come only with his own advent on the scene.[29]

If the difference between conservatism and progressivism lies in the degree of hope that a thinker invests in what could be versus what is

[29] Fourier, *Theory*, 40–44, 56–74.

and has been, Fourier is nearly the ideal progressive. He had as much contempt for the past as perhaps any social theorist who has ever lived. It had, he granted, more by luck and divine providence than by intelligence, produced much of value in the arts and the physical sciences. It had utterly failed to solve the problem—as he claimed to have done, more or less single-handedly—of how to organize society in order to make good use of these discoveries to promote human happiness.[30] Fourier called Francis Bacon to witness to the futility and worthlessness of almost all past philosophizing.[31] The world as the Frenchman saw it was still a hell of suffering for the large majority of its human inhabitants. Incoherent and irrational social institutions maintained by a blind adherence to custom perpetuated misery for the many and furnished happiness only to the fortunate few. If progressivism is measured by a thinker's emphasis on equality and concern for the marginal groups in a society, one again finds Fourier on the emancipationist left. As his best biographer rightly insists, he was no doctrinaire egalitarian. His ideal future included well-defined social classes.[32] Nonetheless, it was a future far more egalitarian than the present and past it would replace. Specific counts in Fourier's indictment of civilization included its vast disparities of wealth and poverty, its subjugation of women, its frequent wars and revolutions, its brutal colonialism, and its plantation slavery.[33]

The key to progress in all of his writings lay in the organization of society along the lines of what he called association. Its basic and largely self-governing unit was to be the community of the phalanstery, though for the many concerns transcending their individual boundaries, there would also be relations among phalansteries, ones both of cooperation and of "societary competition" that would furnish incentives and remuneration for useful inventions in the sciences and the arts.[34] Fourier had no doubt that a single experimental demonstration of a phalanstery in action would make its benefits so clear that all who saw the results

[30] Fourier, *Theory*, 19–23, 101–105.

[31] Fourier, *Theory*, 312; Fourier, *Traité*, I: xii.

[32] Beecher, *Charles Fourier*, 247.

[33] Fourier, *Theory*, 92–95, 104, 129–32.

[34] Fourier, *Theory*, 152–158, 271–272.

would immediately rush to join one. Associated life, to his mind, had two conclusive advantages over the isolated family-household that prevailed under civilization. One was its simple economic superiority. Isolated families were inefficient productive units, and they wastefully duplicated many activities that could much better be done for a large group as a whole. Thanks to the economies of association, members of a phalanstery, Fourier estimated, would enjoy, all told, a standard of living at the very least three times as great as what they could expect outside of one.[35] Who, he asked, could or should resist such a bribe? Greed, to Fourier's mind, was no evil trait that had to be repressed. Under present-day civilization, unfortunately, it led to systematic cheating, speculation, hoarding, and other evils that shortened and embittered the lives of the large majority of human beings. In the era of harmony and phalansterial association, it would impel people irresistibly into the way of life that would make them and others happiest. Fourier expressed nothing but scorn for all past thinkers who had seen the only hope for humankind in the conquest of the passions embedded in human nature. If that were so, he maintained, the situation was indeed hopeless.[36] But it was not. A proper organization of human life could satisfy the passions while at the same time harnessing and usefully exploiting them.

And therein lay the second great advantage of association. The division of labor that was possible in a large community but not in the individual family made possible just such a harmonization of work and pleasure. All that needed doing would be done by those who liked doing it, and so efficiently under association that smaller numbers than had been necessary before would suffice. Women who enjoyed housework would continue to perform it, but those who did not, probably the majority, would find other occupation. The "little hordes" of children would undertake the kinds of work one enjoyed at an early age but later found repugnant. The range of chores open to each individual under association could also satisfy the human passion for variety, and so could the diverse forms of play—for adults as well as children—that association likewise made

[35] Fourier, *Theory*, 160–161; Fourier, *Traité*, I: xv, 46–49.

[36] Fourier, *Theory*, 13–15 (among many other statements of the point).

possible.[37] Fourier rarely let fear of giving offense intimidate him from frankly expounding what he saw as the necessary logic of his reasoning. Some of the more daring extensions that he gave this principle grew out of his unquestioning reverence for all of the human passions. It led him, for example, to the rule of "amorous freedom," which reserved the ties of conventional marriage for the phalansterians who wanted it, but allowed any number of other sexual arrangements for those differently inclined.[38]

As attuned to environmental (though not environmentalist) concerns as any modern Green, Fourier, like Saint-Simon, promised a new earth as well as a new world. Past and present human society, he complained, had done much environmental damage, above all in the greater harshness and unreliability of climate and streamflow resulting from the clearing of the uplands, and he was particularly insistent on the need for mountain reforestation. But he did not believe for a moment as a general rule that nature, in the environmental sense, knew best. His goal was not comprehensive restoration; it was, rather, the utter transformation and improvement of most of the earth's surface apart from the few components of it that happened to be beneficial in their previous or present form.

The technocratic Prometheanism of Saint-Simon and his followers, however ambitious, seems in retrospect rather moderate and realistic. Many of its projects ended up being realized within a century or so. It seems less extravagant than ever when compared with Fourier's program for managing the planet, which shared the fantastically vivid and detailed character of the rest of his plans for the human race. Humankind, he declared, must first increase its numbers and its strength through the adoption of agricultural association. It would then be able to bring under control and cultivation the vast expanses of land presently wild, the domains either of beasts or of human societies so backward as to be little above them.[39] The earth as it existed was beset by climatic and atmospheric disorders and filled with "an immense quantity of harmful animals" and other natural curses: "rattle-snakes, bed-bugs, the legions of insects and reptiles, sea-monsters, poisons, plague, rabies, leprosy, venereal disease,

[37] Fourier, *Theory*, 72–74.

[38] Fourier, *Theory*, 124, 132–143.

[39] Fourier, *Theory*, 47

gout." An earth in heat, its two poles desiring but unable to copulate with each other, continued to bear such harmful and abortive forms of life so long as enlightened human intelligence had not taken it in hand.[40] It had, on the other hand, bred—largely by chance—a few useful creatures, notably the vicuna, the reindeer, the zebra, and the beaver (of which Fourier shared Saint-Simon's high opinion).[41] Humankind's deranged social organization, however, kept it from enjoying their full benefits. Let the human race extend its control over most of the earth's surface, and order and harmony would begin to replace chaos. A warming climate would open additional lands for settlement in the high latitudes, leading to further moderation of temperatures across the earth. It would also melt most of the polar ice and open the long-sought northwest passage for navigation.[42] The tilt of the earth's axis would be scientifically reset to a more appropriate angle.[43] The tropics, like the poles, would become milder and more inviting. Damaging and long-lasting extremes of weather, of "hot or cold, humidity or aridity, storm or calm" would cease to occur.[44] The ocean's chemical composition would be altered, and seawater would acquire the potability and even the taste of lemonade.[45] Sea-monsters, "these vile creatures," would be wiped out by the change and would be replaced by creatures peaceful and useful to man, such as tame new species of whales and sharks. The land surface and the inland waters would also be repopulated by a new and more useful fauna. It could not be said too often, Fourier wrote, that of all conceivable worlds the earth had the most to gain by sweeping away the plants and animals with which previous creations had furnished it and putting new ones in their place.[46] As temperatures grew milder, Fourier supposed, agricultural productions now confined to small parts of the earth would be cultivable over wide areas, and even the worst in quality would be superior to today's best. Much of the transformation of climate would follow more

[40] Fourier, *Theory*, 45

[41] Fourier, *Theory*, 46; Fourier, *Traité* I: 366

[42] Fourier, *Traité*, I: 51–75

[43] Fourier, *Theory*, 49–54

[44] Fourier, *Theory*, 48.

[45] Fourier, *Theory*, 50; Fourier, *Traité*, I: 528–529.

[46] Fourier, *Theory*, 50; Fourier, *Traité*, I: 360–361, 528–529.

or less automatically from "the full exploitation of the globe," the successful cultivation of ever-larger tracts of the earth's surface by a multiplying proliferation of phalansteries.[47]

Fourier pointed out how much of the project of remaking the environment depended on a cooperative organization of society and was impossible under the fragmented, competitive one dear to classical liberal economists. Only collective action, he argued, could successfully eradicate all of the harmful species, from wolves to snakes to crocodiles, from the entire surface of the earth. Only it could undertake the wide-ranging management of water resources necessary to irrigate all of the world's cropland, and only it could drain large morasses whose fragmented ownership precluded such action at present. Only it could make the fisheries of inland waters as productive as they could be by regulating the amounts that individuals could catch.[48] And Fourier envisioned an additional series of conquests of nature requiring the efforts of a dedicated labor force that association alone could make available. Massive public works would be accomplished by traveling battalions recruited from the surplus hands of all phalanxes by what Fourier named the "amorous policy." Young men and women, those physically most capable of the hard work needed, were, fortunately, also those most eager for sexual pleasure, the promise of abundant opportunity for which would draw volunteers for the "industrial armies" traversing and conquering the earth. The efforts of these workers would be further stimulated by the passion of rivalry impelling them to outdo those from other parts of the world laboring by their side on the same projects. Together they would dig ship canals through Suez, Panama, and the landlocked continental interiors, facilitating the movement of people and goods. By planting trees, covering the ground with fertile soil, and modifying the temperature and the winds, they would make wastelands as large as the Sahara Desert productive and prosperous.[49]

Few writers better illustrate than Fourier the risks of misinterpretation connected with the word "nature." If a voluble reverence for nature

[47] Fourier, *Theory*, 274.

[48] Fourier, *Traité*, II: 132–135.

[49] Fourier, *Theory*, 171–178; Fourier, *Traité*, II: 108–113.

in some senses of the word disqualified a thinker as an environmental Promethean, it would disqualify Fourier. But it does not, for he had no reverence at all for physical geography as it stood. Indeed, he had nothing but scorn for those who saw the terrestrial environment as sacred and untouchable, among them "certain scholars who admire spiders, toads, and other such filth."[50] But he accordingly had to reconcile these projects of environmental reconstruction with another of his doctrines: his avowed belief in a wise deity who had created the very earth that needed so much work. He had likewise to explain why a benevolent deity had for so long allowed human beings to welter in social chaos and misery in their ignorance of the way out. He had an answer: God had made the earth as it was, disordered society, oppression of the weak by the strong, horrendous animals, inconvenient isthmuses, brutal climate, and all, as a challenge, in order "to lead us to criticize God," perceive the way that matters ought to be ordered, and "undertake the necessary correctives to his work," beginning with the reconstruction of human society and proceeding to that of the earth. There was no greater impiety, he asserted, than to suppose that the Deity had given the human race and the earth all of the characteristics they possessed without having had good reason to do so.[51]

The most extravagant elements of Fourier's cosmology, such as the promise of a lemonade-flavored ocean, offered easy targets for ridicule. Yet, though his loyal French followers discreetly played down certain elements of the master's thought, they seem to have felt no need to sweep the main elements of his environmental Prometheanism out of sight. On the contrary, they presented its most appealing promises as central elements of his program: deserts and wastelands cultivated and climates made more agreeable by the effort of a properly organized human race.[52]

---

[50] Fourier, *Theory*, 53. He likewise despised wilderness, vastly preferring a tamed and rationalized landscape: Beecher, *Charles Fourier*, 70–71.

[51] Fourier, *Theory*, 53; Fourier, *Traité*, I: 86–87.

[52] See, for example, the anthology of Fourier's work published by his followers: *L'Harmonie universelle et le phalanstère exposés par Fourier: Recueil methodique de morceaux choisis de l'auteur*, 2 vols. (Paris: Librairie phalanstérienne, 1849); also Victor Considérant, *Destinée sociale*, vol. 1 (Paris: Bureau de la Phalange, 1838); Théodore Dezamy, *Code de la communauté* (Paris: Dezamy, 1843); [Henri Dameth], *Notions elementaires de la science sociale de Fourier* (Paris: Librairie de l'école sociétaire, 1844); Hippolyte Renaud, *Solidarité: Vue synthétique sur la doctrine de Ch. Fourier*, 2nd ed.

Fourier observed that Russians must find the climatic changes that his system promised, such as the prospect of seeing orange trees flourish within the Arctic Circle, especially appealing, as they should, too, the promise of a tripling of society's productivity through association.[53] The Petrashevskii Circle, named after its leading figure, which until its forcible suppression in 1849 was the most notable manifestation of political dissidence under the reign of Nicholas I, borrowed its social philosophy chiefly from the French theorist. Though its members were more concerned with society than the environment, Petrashevskii himself proclaimed the advent of an improved and reconstructed earth, and another member put into verse Fourier's promise that the resetting of the earth's tilt would banish winter and summer in favor of eternal spring. A student on the fringe of the circle, N.G. Chernyshevsky (1828–1889), escaped its breakup to emerge as the leading publicist of Russian radicalism in the freer days that followed Nicholas's death in 1855. He wrote a utopian novel, *Chto delat'?* (1863) (What Is to Be Done?) while in jail himself awaiting trial on charges of sedition. The novel, though amateurishly written, became hugely influential among the educated Russian left during Chernyshevsky's long Siberian exile. Fourierist in inspiration, it depicted the lives of a set of principled young radicals who form a communal establishment on rigorously egalitarian lines. In the dreams of one of the characters, Vera Pavlovna, it offered as well a recognizably Fourierist vision of a progressive future world whose progress in conquering the planet's environment is as notable as its parallel reform of the relations of human beings to one another. It is eliminating the marshes and deserts that defaced the earth in the past, the one by drainage, the other by irrigation, cultivation, and climatic improvement. It is conquering mountains, too, that were formerly only "barren cliffs," and covering them with productive gardens.[54]

(Paris: Librairie sociétaire, 1845); and Just Muiron, *Transactions sociales*, 2nd ed. (Paris: Librairie sociétaire, 1860).

[53] Fourier, *Traité*, II: 641.

[54] V.E. Evgrafov, ed., *Filosokskie i obshchestvenno-politicheskie proizvedeniia Petrashevtsev* (Moscow: Gosudarstvennoe izdatel'stvo politicheskoi literatury, 1953), 427–428, 428n, 692; Nikolai Chernyshevsky, *What Is to Be Done?*, trans. Michael R. Katz (Ithaca, NY: Cornell University Press, 1989), 373, 374–375, 377.

The USA proved more hospitable to the theory of association than any other country besides France. American enthusiasts set up several dozen phalansteries, though hardly any of them lasted for long, and the existing New England utopian community of Brook Farm converted to the new doctrine.[55] Elements of Fourier's thought diffused into the work of mainstream writers, chiefly of a progressive persuasion, from the journals that his followers established and the books that they wrote. Reworking the *Théorie* and the *Traité* into a volume for American consumption, the New York journalist Albert Brisbane bowdlerized them of some of Fourier's more unsettling ideas about social relations, particularly those between the sexes.[56] All the same, he faithfully passed along and elaborated the master's hopes for the remaking of the global environment. Brisbane and his fellow American Fourierists promised that the conquest of the earth by the associated activity of the phalanstery would create "serene and genial climates." Extremes of weather would disappear once association transformed the earth's surface, and so would other features that defaced it: "its vast forests, its unregulated streams, which overflow and ravage the plains in their course, its stagnant waters, and sandy deserts." Disease too would cease once "a perfect cultivation extended over the whole earth, and all nature were brought into a state of order." But "those great labors … which are necessary to bring the earth and the atmosphere into a healthy condition, are impossible," they warned, so long as society remained fragmented and competitive. Communitarian socialism alone could carry them out.[57] Enjoying for some time the hospitality of the reformist editor Horace Greeley on the front page of his fledgling *New York Tribune*, Brisbane described for its readers in 1842 what Fourierism meant in environmental matters:

---

[55] Carl Guarneri, *The Utopian Alternative: Fourierism in Nineteenth-Century America* (Ithaca, NY: Cornell University Press, 1991), 407–408.

[56] Ibid., 93–98.

[57] *The Phalanx*, vol. 1 (1844), 312; *The Harbinger*, vol. 2 (1846), 194; vol. 4 (1847), 121; vol. 1 (1845), 174. See also, e.g., Albert Brisbane, *Social Destiny of Man; or, Association and Reorganization of Industry* (Philadelphia, PA: C. F. Stollmeyer, 1840), 73, 239–240, 244–245, 263–267 and Parke Godwin, *A Popular View of the Doctrines of Charles Fourier* (New York, NY: J. S. Redfield, 1844), 78–84, 104.

> Man … must cultivate and embellish his Globe, clear it it [sic] of its rude forests, fertilize its wild plains and valleys, dike and regulate its streams, drain its marshes, reclaim its deserts, develope its varied material riches and resources, efface Discords and establish Harmonies, and improve and adorn it by every means which his Genius can suggest and his Labor execute … let them unite and assemble, and they could take this Globe,—now so much neglected and devastated, with its burning Deserts, its pestilential Marshes, its rude Forests, its uncultivated Plains, its bleak and barren ranges of Mountains, and its turbid Streams, and … they could cultivate and embellish it, until they transform it into a magnificent terrestrial Abode, worthy of the God who created it, and the Genius of Man which presides over it.[58]

Greeley himself endorsed the associationist vision of "renewing and beautifying the earth, until at last even the Polar Ices should be dissolved, and a joyous, exhilarating spring-time envelop our planet. The reclamation of deserts, of pestilential marshes, of wildernesses and snow-capped mountains … all these, and many more dizzying, are among the ultimate consequences of Social Reorganization anticipated by Fourier." No more than the French theorist did Greeley, or any Promethean, accept any and all human modification as necessarily an improvement; like Fourier, for instance, he called for the reforestation of uplands that had been unwisely cleared of trees. But his bias too was strongly toward nature improved by human action, and it extended to the global scale Ignatius Donnelly's program for conquering the arid American West. "I have a firm faith," Greeley stated near the end of his life, "that all the great deserts of the Temperate and Torrid zones will yet be reclaimed by irrigation and tree-planting." A modern historian has observed: "The subjugation of nature, which has come to seem to many in our environmentally conscious day the very opposite of reform, seemed to Greeley (and many other Americans of his day) synonymous with it."[59]

[58] Albert Brisbane, "What Is the Destiny of Man?," *New-York Daily Tribune*, April 22, 1842, 1.

[59] Horace Greeley, *Hints Toward Reforms, in Lectures, Addresses, and Other Writings* (New York: Harper & Brothers, 1850), 292; Horace Greeley, *What I Know of Farming* (New York: G. W. Carleton & Co., 1871), 46–47, 51–52, 137, Daniel Walker Howe, *The Political Culture of the American Whigs* (Chicago, IL: University of Chicago Press, 1979), 188. On Greeley and socialism,

The American communal socialist press, and Greeley's *Tribune* as well, warmly welcomed another prophet of environmental transformation, the German immigrant and Hegelian socialist John Adolphus Etzler (1791-?). In his writings, beginning with his book *The Paradise Within the Reach of All Men* (1833), Etzler proposed to harness, regulate, and apply four superabundant yet so far little-exploited sources of power—the winds, the waves, the tides, and the sun—to a Promethean remaking of the earth's surface and the creation of a just and equal society with material abundance for all.[60] His reliance on what would today be called renewable energy sources did not by any means make him an environmentalist, for the gist of his argument was that they offered a sufficient means "to change the whole face of nature, on land and sea, into whatever man wishes to see, by a magic like power." With it, one could at will "level mountains, sink valleys, create lakes, drain lakes and swamps, intersect everywhere the land with beautiful canals," build roads, irrigate deserts, create soil, reclaim coastlands, prevent rivers from flooding, make the oceans as productive as the land, explore and exploit the planet's interior, construct marvelous new buildings and grounds for human habitation, and make travel from one part of the world to another easy, rapid, and safe. Etzler dwelt with especial relish on the ease with which forests could henceforth be cleared for farmland: "Any wilderness, even the most hideous and the most sterile, may be converted into the most fertile and delightful gardens." An earth so transformed, he predicted, would easily supply the wants of a population a thousand times its current size.[61] He fashioned a versatile machine, which he called a "Satellite," that would be capable of performing all the major tasks of engineering the land surface, able to "eradicate trees, remove rocks, saw the trees into pieces on the spot where they grew, till the land in the most superior manner as

see Adam Tuchinsky, *Horace Greeley's New-York Tribune: Civil War-Era Socialism and the Crisis of Free Labor* (Ithaca, NY: Cornell University Press, 2009).

[60] *New-York Tribune*, 16 July 1841, 4; *New-York Daily Tribune*, 4 November 1843, 8 and 4 December 1843, 1; John A. Etzler, *The Paradise Within the Reach of All Men* (Pittsburgh, PA: Etzler and Reinhold, 1833), 118. The fullest account of Etzler's career is Steven Stoll, *The Great Delusion: A Mad Inventor, Death in the Tropics, and the Utopian Origins of Economic Growth* (New York, NY: Hill & Wang, 2008).

[61] Etzler, *Paradise*, 1, 68, 118; John A. Etzler, *Emigration to the Tropical World, for the Melioration of All Classes of Men of All Nations* (Ham Common, Surrey, 1844), 5.

a garden … with irrigation or draining of the soil wherever it may be desirable; make roads and canals."[62] His environmental utopia was intimately tied to a social, and, indeed, socialist one. Only proper principles of human association, not those of competitive capitalism, Etzler maintained throughout his writings, could bring such forces to bear in the vast labor of remaking the earth, and in turn the abundance that they created would, by abolishing the scarcity that lay at the root of poverty, oppression, ignorance, and inequality, sweep away a host of evils, from famine to slavery to the subjection of women to commercial fraud to war.

The prickly libertarian and ardent naturalist Henry David Thoreau gave Etzler's work its best-remembered review, a mocking and disparaging one.[63] The American Fourierists, on the other hand, expressed only enthusiasm for the help that the new technology would provide in the work of "refining and perfecting nature … the control of vast mechanical power will confer on [man] more unlimited dominion and render him more emphatically what he was designed to be by the Creator, King of the Earth." In their "Dialogue on Etzler's Paradise," a Fourierist named "Clear" explained to "Dunce" and "Flat" the potential of the German's inventions to "remove mountains" and "level a hill and carry it into the next valley in a single night." Etzler won their praise not merely for providing the world with an improved means for "digging canals, draining large swamps, clearing forests, &c." but also for understanding the essential link between such activities and the reform of society.[64]

Across the Atlantic, Etzler's proposals won an equally warm reception from the small circle of British Fourierists in their journal *The London Phalanx*, and likewise from the disciples of Robert Owen, the third member (with Saint-Simon and Fourier) of the classic trinity of early nineteenth-century European utopian socialism.[65] Owen shared his

---

[62] Ibid., 17.

[63] Henry David Thoreau, "Paradise (To Be) Regained," in *Collected Essays and Poems*, ed. Elizabeth Hall Witherell (New York, NY: Library of America, 2001), 115–137.

[64] *The Harbinger* 1 (1845), 383–384; *The Phalanx*, 1 (1843–44), 30, 271–273.

[65] *The London Phalanx*, #24 (1841), 375–376; Gregory Claeys, "John Adolphus Etzler, Technological Utopianism, and British Socialism: The Tropical Emigration Society's Venezuelan Mission and Its Social Context, 1833–1848," *English Historical Review*, vol. 101, #399 (1986), 351–375; George Jacob Holyoake, *The History of Co-Operation in England: Its Literature and Its Advocates*, vol. 1 (London: Trubner & Co., 1875), 213–214.

compeers' emphasis on the unity of social and environmental progress. He was closer in the substance of his diagnosis and remedy to Fourier than to Saint-Simon—though Fourier attacked both of them as charlatans and false guides who would lead the world astray from his own one correct path.[66] Owen, too, recommended the creation of communal associations, approximately the same size as Fourierite phalansteries, as the necessary form of utopian life, and he treated such established social institutions as marriage, orthodox religion, and private property with the same disrespectful freedom as his French counterpart. Eschewing Fourier's elaborate periodization, though, Owen reduced world history to a contrast even more simple and stark than Saint-Simon's chronology. The past and present, for Owen, were the dominion of misery and ignorance. The future, once dysfunctional social institutions disappeared and "the change from irrationality to rationality" had occurred, would see the universal reign of harmony and happiness.[67]

Owen, like many radicals of his day, regarded the Malthusian thesis of natural limits to human well-being as one of the great intellectual props of the evil status quo.[68] He flatly rejected the thesis that had become synonymous with Malthus's name, that growth in population would absorb and nullify any advances in wealth and production that the world might achieve. On the contrary, he argued, the progress of science and invention was outstripping, and would continue to outstrip, demographic increase. Prosperity would rise as the number of people did. Demographic increase, moreover, would furnish the labor that would for the first time make the human mastery of the entire planet's surface possible. Owen challenged his opponents to say whether they did not agree with him that "the whole earth should be fertilized, and made healthy and beautiful, in the shortest time practicable," and that the scarcity of human hands to perform the work was the chief obstacle to attaining that goal. At present, he asserted, "the earth is comparatively a waste and a forest for want of people to

---

[66] Charles Fourier, *Pièges et charlatanisme des deux sectes, Saint-Simon et Owen, qui promettent l'association et le progrès* (Paris: Bossange père, 1831).

[67] On Owen and his movement, see J.F.C. Harrison, *Robert Owen and the Owenites in Britain and America: The Quest for the New Moral World* (London: Routledge and Kegan Paul, 1969).

[68] For Owen's succinct dismissal of Malthus, see his *A New View of Society; or, Essays on the Formation of the Human Character*, 3rd edition (London: Longmans et al., 1817), 174–176.

drain and cultivate it," far too much of its surface "wild, barren, waste, swamp, and forest." Should humankind adopt his system of society: "The earth, as population can be made to increase, will be reclaimed from its original wild condition, drained, fertilised, made healthy and beautiful." "Illimitable, innumerable, new powers from science have been given," Owen added in a classically Promethean vein, "to aid man's limited natural power, to enable him to become an active agent in making our planet a terrestrial paradise."[69]

Owen was an important figure in British religious freethought as well as in British socialism. Some of his fellow unbelievers in subsequent years adhered more or less formally to the Positivism of Saint-Simon's breakaway disciple Auguste Comte, who proposed a "Religion of Humanity" to replace the supernatural creeds of the past. In his positivist rhapsody *The Martyrdom of Man* (1872), Winwood Reade foresaw science and invention conquering the earth's surface and making its resources superabundant, overcoming disease and then death and furnishing the technology of space travel to make human beings first colonists of the rest of the universe and then "manufacturers of worlds."[70] The foremost spokesman of organized English Positivism, Frederic Harrison, observed that the poets of nature and what he called the "nature-worshippers" took a very narrow and partial view of the terrestrial environment. They ignored its abundant and appalling dark side. "Waste, ruin, conflict, rot are about us everywhere" in nature, he pointed out; "animal, and vegetable, and inorganic life are at war, tearing each the other; every leaf holds its destructive insect; every tree is a scene of torture, combat, death; everything preys on everything; animals, storm, sun, and snows waste the flower and the herbs; climate tortures to death the living world, and the inanimate world is wasted by the animate, or by its own pent-up forces." Even the beauties the nature-worshippers celebrated were largely the work of man, both physically—"The earth was a grisly wilderness till man appeared; and it

[69] Robert Owen, *A Development of the Principles and Plans on Which to Establish Self-Supporting Home Colonies* (London: Home Colonization Society, 1841), vi, 40; Robert Owen, *The Revolution in the Mind and Practice of the Human Race; or, The Coming Change from Irrationality to Rationality* (London: Effingham Wilson, 1849), 94–95, 122.

[70] Winwood Reade, *The Martyrdom of Man* (London: Trübner, 1872), 511–515 (quotation from 515).

had but patches of beauty here and there, until after man had conquered it"—and intellectually. When readers thought that they were admiring the beauties of nature, Harrison suggested, they were really admiring the genius of the poets who, by the power and selectivity of their vision, had created those beauties.[71]

In writing thus, Harrison echoed Auguste Comte himself. Poetry, the founding Positivist declared in the 1840s, had not caught up with the march of events. It still imitated the literature of antiquity in making nature, despite its "extreme imperfection," the principal object of its admiration. But in modern times, Comte asserted, whatever had been the case in classical times, what was made by human hands and minds was plainly superior in every way to what was made by nature. Poetry must find its inspiration henceforth in "the prodigies of man, his conquest of nature," and in doing so it would win a greater popularity than ever, for it would at last be in harmony with human beings' deep and justified sense of their own superiority over their environment. In that same spirit, Comte described man as "the supreme governor of nature's economy, which he ceaselessly modifies to his own advantage" while feeling—and rightly feeling, Comte believed—no scruples about doing so.[72] He made a particular point of justifying the eradication of those species for which humankind could find no useful purpose, along with the alteration and improvement of the rest.[73] He made the earth one of the sacred "fetishes" of his Religion of Humanity, but it was the earth as reshaped by his supreme fetish, the united human race itself.[74]

Outright Promethean atheism had a plausible basis, for what remained of unconquered nature furnished disbelievers in God, or at least in Christianity, with abundant evidence of shoddy workmanship by the

[71] Frederic Harrison, *The Creed of a Layman: Apologia pro fide mea* (London: Macmillan, 1907), 200–201.

[72] Auguste Comte, *Cours de philosophie positive*, vol. 6 (Paris: Bachelier, 1842), 881–882.

[73] August Comte, *Système de politique positive*, vol. 1 (Paris: L. Mathias, 1851), 614–619; vol. 2 (Paris: L. Mathias, 1851), 72.

[74] On Comte's Prometheanism, see also Jean-Francois Braunstein, "Comte, de la nature à l'humanité," in Olivier Bloch, ed., *Philosophies de la nature: Actes du colloque tenu à l'université de Paris* (Paris: Publications de la Sorbonne, 2000), 259–269 and Andrew Wernick, *Auguste Comte and the Religion of Humanity: The Post-Theistic Program of French Social Theory* (Cambridge: Cambridge University Press, 2001), 165–177.

purported creator. As it had done for Shelley, the evidence of an imperfectly designed world provided arguments simultaneously against orthodox Christianity's omniscient maker of the earth and for human efforts to correct the flaws of physical geography: first and foremost, those of the weather. The celebrated American freethinker Robert G. Ingersoll asked his opponents:

> Do you not believe that any honest man of average intelligence, having absolute control of the rain, could do vastly better than is being done? Certainly there would be no droughts or floods; the crops would not be permitted to wither and die while rain was being wasted in the sea. Is it conceivable that a good man with power to control the winds would not prevent cyclones? Would you not rather trust a wise and honest man with lightning?[75]

They could, of course, have replied that the earth was a fallen world whose nature, like human nature, had suffered and required redemption. But for all that it had a warrant ready-made in the celebrated Biblical injunction to subdue the earth, Prometheanism aroused little enthusiasm among believers in conventional forms of Christianity, and for reasons that are not difficult to understand. Established doctrine inclined believers to look for a perfected environment in the world to come and not on earth, or else to expect a redeemed and perfected earth, not through human effort tinged with a flavor of rebellion against the government of the universe—one of the features of the myth of Prometheus—but bestowed as a divine gift. Full-blown Prometheanism carried a suggestion of impiety in presuming to improve upon the design of the Creator. When found among the religious, it was chiefly in marginal settings, as witness the chilly reception of Teilhard de Chardin's noospheric speculations by his Roman Catholic superiors.

The most important Western form of unorthodox theistic belief around the end of the eighteenth century and the beginning of the nineteenth was Deism or natural religion. Sharing conventional Christianity's belief in a wise and benevolent Creator, it parted company on a number

[75] Robert G. Ingersoll, "A Reply to the Rev. Henry M. Field, D.D.," *The North American Review* 145, #372 (1887), 479.

of further points. Man was good by nature, Deists held, and as he had not fallen, neither had the earth. The Creator did not intervene actively in the creation through such modes as miracles, special providences, or revelations. Such a creed might seem likelier to discourage than to promote a Promethean program. How could human beings improve in any way upon an earth fashioned by a divinity at once omniscient and benevolent? But impulses less logical than psychological carried natural religion in that direction nonetheless. Distancing the Creator from the world, as it did, dispelled something of the aura of impiety and presumption previously hovering around the idea of large-scale environmental engineering. One could also suppose, as Fourier would do, that God had given man reason in order to see what tasks needed doing, or that a physical geography suited to the infancy of human society must be changed as that society grew to maturity. Finally, Deism's exaltation of reason and of scientific inquiry into nature's design led some distance, however accidentally, toward a heightened faith in the power of reason and science to redesign what they found.

In any case, the greatest American enthusiast for the remaking of the earth by human hands in the early national period was a devotee of natural religion, the poet-diplomat Joel Barlow (1754–1812).[76] In political, religious, and environmental matters alike, he put more stock in the future than in the past. An enthusiast both for the American Revolution, in which he fought, and for the French Revolution, which he witnessed in person and encouraged with his pen, Barlow was an outcast from the circle of Connecticut poets and public men to which he had belonged as a young man, but whose conservative Federalism he rejected to ally himself politically with the northern followers of Thomas Jefferson. Barlow's writings coupled the political and social progress that he understood chiefly as humankind's liberation from the fetters of tradition with its ever-growing mastery over the surface of the planet. He fused the two in what he regarded as his most important work, a national epic poem for the new USA, first published as *The Vision of Columbus* (1787) and enlarged into *The Columbiad* (1807).

---

[76]On Barlow's life, see Richard Buel, *Joel Barlow: American Citizen in a Revolutionary World* (Baltimore, MD: Johns Hopkins University Press, 2011).

The literary scholar Cecelia Tichi has described Barlow's environmental ideal that of "the engineered millennium."[77] His enthusiasm for the remaking of the earth's surface pervaded his verse and overlapped with his politics. With his friend Robert Fulton, he agitated for the construction of a network of canals and river improvements to overcome the obstacles that unreformed physical geography opposed to communication within the USA and elsewhere. As a conquest of enlightened reason, Tichi wrote, such a network was for Barlow and his associates a "mystical symbol" and not merely an aid to transportation; it stood for the mastery of nature, which itself represented the social forces that oppressed humanity.[78] Barlow's hoped-for environmental reforms, all carrying the same ideological charge, likewise included the clearance of the encumbering forests, the comprehensive drainage of wetlands for cultivation and settlement, the irrigation of farmlands that needed it, the diking of great rivers to protect their banks from floods, and the opening of passages for shipping through the barriers of Darien and Suez.[79] He envisioned a future in which dangerous storms on sea or land would be tamed, in which the ability to make rain by artificial means would dispel all of the terrors of drought, in which the clearing and cultivation of the land would moderate its temperatures and banish the severity of American winters.[80] He foresaw a comprehensive mastery by man of the terrestrial surface: "Labors that fasten to his sovereign sway/Earth's total powers, her soil and air and sea/Force them to yield their fruits at his known call/And bear his mandates round the rolling ball."[81]

Barlow consistently mingled these expectations with ones of political and social transformation that he saw as their necessary complements. Only the liberation of the human mind from the bonds of tradition, the

[77] Cecelia Tichi, *New World, New Earth: Environmental Reform in American Literature from the Puritans through Whitman* (New Haven, CT: Yale University Press, 1979), chapter 4, "Joel Barlow and the Engineered Millennium," 114–150, is an excellent analysis of his environmental visions and the social theories that he linked to them.

[78] Ibid., 146–148 (quotation from 147); Kenneth R. Ball, "Joel Barlow's "Canal" and Natural Religion," *Eighteenth-Century Studies* 2, #3 (1969), 225–239.

[79] Joel Barlow, *The Columbiad: A Poem, with the Last Corrections of the Author* (Washington, DC: Joseph Mulligan, 1825), 35, 281–282, 342–345.

[80] Ibid., 46, 194–200, 346–347, 358, 369–370.

[81] Ibid., *Columbiad*, 360.

free advance of science, and the replacement of tyrannies and aristocracies by democratic republics could bring them about. Other Americans of his time duplicated Barlow's fusion of natural religion, human progress, and Promethean engineering. One distinguished Jeffersonian scientist wrote in an open letter to another contrasting modern research into the weather with the beliefs of old times that attributed it and its variations directly to the will of God, and he concluded with the hope that such research might help to "form a system which may teach us to bridle the *winds* themselves, and render them farther subservient to human uses."[82]

A kindred figure to Barlow in English thought was Erasmus Darwin (1731–1802), best remembered today as Charles Darwin's grandfather, but famous in his own time as poet, physician, inventor, and political radical. In his social and environmental views, Darwin had much in common with Shelley, who would become an admirer of his work. Religiously unorthodox, often classified as a Deist but at the same time a rather un-Deist-like believer in the evolution rather than the static design of the universe, Darwin expected human intelligence to play an increasingly important role, through the application of technology, in its further development. His epic poems celebrated the transformation of the landscape by the application of science, notably the works of his friend the engineer and canal builder James Brindley. They also predicted triumphs of technology that included the control of the weather. Humankind, Darwin speculated, might develop the means of redirecting the great currents of the atmosphere in order to improve the distribution of climates across the earth's face. Or it might warm the polar regions and cool the equator by maneuvering icebergs in large numbers toward the lower latitudes.[83] Politically and theologically conservative critics on both sides of the Atlantic mocked such dreams as utopian, mingling ridicule for their impracticality with distaste for their impiety.[84]

---

[82] Benjamin Waterhouse, "Observations and Reflections on Storms, and Other Phenomena of the Atmosphere," *Medical Repository*, 5 (1802), 472.

[83] Desmond King-Hele, *Erasmus Darwin: A Life of Unequalled Achievement*, (London: Giles de la Mare, 1999), 227–228, 260; Maureen McNeil, *Under the Banner of Science: Erasmus Darwin and His Age* (Manchester: Manchester University Press, 1987), 20–21, 78–79, 90–91.

[84] *Mercury and New-England Palladium*, 15 May 1801, 1; *New-England Palladium*, 13 October 1801, 3; *Boston Gazette*, 30 May 1805, 1; Benjamin Silliman, *Travels in England, Holland, and Scotland*, 2nd ed. (Boston, MA: T. B. Wait, 1812), I: 20; [Thomas Green Fessenden], *Terrible*

A poet whose epic ambitions equaled Barlow's and Darwin's and whose achievements far exceeded theirs was France's Victor Hugo (1802–1885). His religious beliefs, too, have usually been described as Deistic, a label that fits as well as any other word, though they were too idiosyncratic to fall neatly into any ready-made category. In more earthly matters, Hugo stood to France, and, indeed, the world, as incarnating the ideals of progress and equality, ones that he coupled with an admiration for inventors and engineers and with visions of nature as humankind's great adversary and of its conquest through the marvels of technology.[85] His dreams of progress included "clearing the globe, colonizing the deserts, improving the creation under the Creator's eye," as he proposed in 1849. In a speech three decades later, Hugo pointed to such huge engineering projects as Lesseps's canals and Roudaire's North African inland sea as realizations of his hopes. "Man is becoming master of the earth," he declared. "Do you want to pierce an isthmus? You have Lesseps. Do you want to create a sea? You have Roudaire. … Do you need a sea? Create it, and it will create navigation, and navigation will create cities. To whomever wants a field, say: Take it. The earth is yours, cultivate it."[86]

The French humanitarian socialist Pierre Leroux (1797–1871), Hugo's contemporary and fellow exile during the Second Empire, also espoused a natural religion that laid a heavy emphasis on the role of human initiative in realizing both "the perfectibility of man and the perfectibility of nature."[87] It was humankind's appointed task, Leroux wrote, to complete God's rough first sketch of the earth's surface. As it did so, wild forests and brush gave way to fields and pastures; rivers submissively flowed in new beds; "the wayward floods of the plain are confined within banks … new waterways appear and furrow the ground with their basins and

Tractorations! in Four Cantos, by Christopher Caustick, 2nd American edition (Boston, MA: Lorenzo Press of E. Bronson, 1806), 51, 98; [T. J. Mathias], *The Pursuits of Literature: A Satirical Poem in Four Dialogues, with Notes*, 7th edition (London: T. Becket, 1798), 115, 210n; "The Pneumatic Revellers: An Eclogue," *The Anti-Jacobin Review and Magazine* 6 (1800), 109–110.

[85] Katherine Lunn-Rockliffe, "Humanity's Struggle with Nature in Victor Hugo's Poetry of Progress," *Modern Language Review* 107, #1 (2012), 143–161.

[86] Victor Hugo, *Actes et paroles, I: Avant l'exil, 1841–1851* (Paris: Michel Lévy frères, 1875), 383–384; *Actes et paroles III: Depuis l'exil, 1876–1885* (Paris: Albin Michel, 1940), 483.

[87] Pierre Leroux, "Refutation de l'eclecticisme," in *Oeuvres de Pierre Leroux (1825–1850)*, vol. 2 (Paris: Louis Nétré, 1851), 313.

canals; the mountains are flattened," and even the nighttime darkness of the planet was being banished by the lights that human effort kindled.[88] An acerbic critic of Malthus, Leroux reversed the Englishman's theorem; subsistence, he wrote, was not finite but potentially infinite, increasing with population, and he quoted against Malthus the Biblical command to be fruitful and multiply, and fill the earth.[89] He was convinced that the use of human wastes for fertilizer would close the circle between production and consumption and permit the soil to sustain permanently any number of human inhabitants. His vision, if implemented, would have covered much of the earth's surface with a landscape of intensively engineered agriculture, served by an elaborate network of pipelines for sewage irrigation and for drainage. Large areas of idle wasteland currently uncultivated for want of fertilizer would thus be brought into use. Land in its natural condition Leroux thought wastefully unproductive and far short of its potential under human management, nor, he argued, did people need to rely on the slow processes of nature to create the fertile land that they could now make for themselves.[90] Like Fourier, he illustrates how easily an expressed reverence for a frequently invoked "nature" would go along with Promethean plans to transform the earth's surface.

Leroux's onetime collaborator Jean Reynaud (1806–1863) exercised much influence on French socialist thought in his day.[91] In his most ambitious work, *Terre et Ciel* (1854), Reynaud preached a theory of reincarnation and of the migration of spirits through successive existences among a plurality of inhabited worlds, but his concern with the wider universe did

[88] Pierre Leroux, "Trois discours: Aux artistes," in *Oeuvres de Pierre Leroux (1825–1850)*, vol. 1 (Paris: Société typographique, 1850), 63–64.

[89] Pierre Leroux, *Malthus et les économistes, ou: Y aurait-il des pauvres?*, 2nd edition (Boussac: P. Leroux, 1849), 89, 184–89; Pierre Leroux, *Aux états de Jersey: Sur un moyen de quintupler, pour ne pas dire plus, la production agricole du pays* (London: Universal Library/Jersey: L. Nétré, 1853), 9.

[90] Ibid., 26, 30, 37; Pierre Leroux, *De la ploutocratie; ou du gouvernement des riches*, 2nd edition (Boussac: P. Leroux, 1848), 153. On this element of his thought, see Ceri Crossley, "Pierre Leroux and the *Circulus*: Soil, Socialism and Salvation in Nineteenth-Century France," in Louise Lyle and David McCallam, eds., *Histoires de la Terre: Earth Sciences and French Culture, 1740–1940* (Amsterdam: Rodopi, 2008), 105–118.

[91] On his life, thought, and influence, see Guillaume Cughet, "Utopie et religion au XIXe siècle: L'oeuvre de Jean Reynaud (1806–1863), théologien et saint-simonien," *Revue historique* 306, #3 (2004), 577–599.

not preclude a close attention to the problems of life on earth. First and foremost among those problems he placed humankind's battle against the indifference or downright hostility of natural features and processes that showed no signs of having been designed for its welfare. "Industry," a term that he, like Leroux, had picked up as a young Saint-Simonian, had a truly religious character for Reynaud; it meant conquering nature to make human life more physically secure and materially richer. If all changes wrought in the earth by people were suddenly erased and nature in its original state restored, he wrote, "what a frightful calamity" it would be! The forests would return at the expense of cultivated land, much of which would revert to its "natural sterility," the supply of food would diminish drastically, and the useless species that long effort had eradicated or controlled would proliferate, while the useful ones would lose all of the improvements that their long domestication had developed. Only ceaseless effort, a "warfare against nature" carried on by force and violence, Reynaud proposed, had maintained and could continue to maintain any of the changes for the better that past effort had brought about. Such warfare, he continued, represented the principal task of the human race. But the heavy burden of labor it imposed could only become lighter as science developed. There was no limit to the progress of inventions able to correct nature's faults, nor to the possibilities for turning its titanic forces to the task of its own transformation. The energies of the winds, the rivers, the tides, the earth's internal heat, and perhaps even its electricity "all belong to us, if we wish … what could we not do with them if we had them at our command?," as he believed the future would.[92]

No less concerned than Reynaud with terrestrial matters, for all the similar otherworldiness of his concerns, and no less associated with political and social radicalism, was the leading American spiritualist of the nineteenth century, Andrew Jackson Davis (1826–1910). Along with a commitment to social reform of a progressive, egalitarian, and communitarian kind, Davis proposed as vast a scope for the physical improvement of the earth as any environmental Promethean of his day. He denied that there was any irreverence in improving a planet that God had created. As

[92] Jean Reynaud, *Terre et ciel: Philosophie religieuse*, 2nd edition (Paris: Furne, 1854), 93–94, 97, 98, 103–104.

Fourier had done—indeed, he may have drawn the argument from the French writer—Davis argued that just such alteration was humankind's assigned work on earth. "Did the Deity do anything for man, which man can, by social progress and intellectual development, accomplish for himself? Far from it." The things left to be accomplished, in his estimation, included the clearing of forests, the transformation of swamps and deserts into fertile farmland, the elimination of "the disagreeable vegetables and animal forms, which now disfigure the face of nature," and a control over the atmosphere that would melt the polar ice, opening new routes for navigation and new lands for settlement, and permit rain to be made at will.[93]

No one in the period offered a more ambitious Promethean program or embedded it in a more comprehensively worked-out philosophy and theology than the nominally Christian but highly unorthodox thinker Nikolai Fedorovich Fedorov (1829–1903). Born in provincial southern Russia, the illegitimate son of a well-to-do noble landowner, Fedorov claimed to have conceived as early as 1851 the central ideas that he would devote the rest of his life to working out. After teaching history and geography in a succession of provincial cities, he held the post of librarian of Moscow's Rumiantsev Museum from 1874 to 1898, exchanging it 1898 for a similar position in the Ministry of Foreign Affairs. He led an ascetic life, but one filled with lively interaction with a circle of friends and with library patrons, whom he was assiduous in helping pursue their studies. Largely avoiding publication under his own name, he dictated or inspired a number of anonymous articles expounding his ideas, which also circulated widely by word of mouth. After Fedorov's death in 1903, they began to reach a wider public through the two-volume collection *Filosofiia obshchego dela* (*The Philosophy of the Common Task*), assembled,

[93] Andrew Jackson Davis, *Harmonial Man; or, Thoughts for the Age*, 4th edition (Boston, MA: Bela Marsh, 1856), 25–35 (quotations from 27 and 32); also 35–80 for Davis's detailed proposals on how to make rain. See also *The Penetralia; Being Harmonial Answers to Important Questions*, revised edition (Boston, MA: Colby & Rich, 1872), 350–351. On Davis's politics, see Robert W. Delp, "Andrew Jackson Davis and Spiritualism," in Arthur Wrobel, ed., *Pseudo-Science and Society in Nineteenth-Century America* (Lexington, KY: University Press of Kentucky, 1987), 100–121.

published, and distributed by two disciples in small editions in 1906 and 1913.[94]

As instrumentalist as John Dewey, just as disparaging of research pursued for the sake of idle curiosity, Fedorov valued thought and theory only as guides to action. His "common task" would transform what he called science as the search for facts into science as a purposeful project, the project of taking control of and guiding the future: alike that of humankind, the earth, and the universe.[95] He grounded the task in Christian doctrine, but in a highly unorthodox way, seeing the resurrection of the dead to immortal life, both in body and soul, as something to be achieved by humankind itself, not as a divine gift to be bestowed in another world. Only by consciously recognizing their obligation to undertake this common task and voluntarily uniting in it, he argued, could human beings successfully attempt the work of resurrection, and only by entirely remaking the world of nature, both on earth and beyond, could they fulfill it. For, at present, nature (both terrestrial and cosmic) was a realm of mindless physical forces. In carrying out Fedorov's common task, human beings (physically weak, but gifted with reason) would reconstruct the material world (physically strong, but bereft of mind or purpose), redirecting its blind energies to beneficent and purposeful ends. Simply in order to secure its own existence, as Fedorov saw it, the human race would eventually have to colonize and rebuild the cosmos itself. Its first task was to bring the natural forces at work on the earth's surface under control, in particular those governing the weather and climate.[96]

Though he spoke of it often as an enemy, Fedorov did not regard uncontrolled nature as malevolently disposed toward human beings. He saw it, rather, as a chaotic realm of mindless elemental powers whose

---

[94] The fullest account of Fedorov's life in English is in George M. Young, *The Russian Cosmists: The Esoteric Futurism of Nikolai Fedorov and His Followers* (New York, NY: Oxford University Press, 2012), 46–75. A scholarly edition of his writings appeared between 1995 and 2000, edited by A. G. Gacheva and S. G. Semenova: *N. F. Fedorov: Sobranie sochinenii v chetyrekh tomakh*: v. 1, Moscow: Progress, 1995; v. 2, Moscow, Moscow: Progress, 1995; v. 3, Moscow: Traditsiia, 1997; v. 4, Moscow: Traditsiia, 1999; and [v. 5] (additions and notes to vol. 4): Moscow: Traditsiia, 2000. References below to this edition (Fedorov, *Sobranie sochinenii*) are keyed to these volume numbers.

[95] See, e.g., Fedorov, *Sobranie sochinenii*, II: 240–241.

[96] Ibid., I: 121.

chance workings were sometimes beneficial, and just as often disruptive and "death-dealing." Nature, to him, was at once the source of life and of death, eventually destroying everything that it created.[97] The great drought and subsequent famine that devastated southern Russia in 1891–1992 stimulated much of Fedorov's environmental thought and served him as a particularly powerful example. Humanity's great challenge, he argued, was to transform the natural processes that could produce such calamities into permanently beneficial agencies under rational control.

No more than any intelligent Promethean did Fedorov approve of any and all human alterations of nature, but he was much farther still from urging its preservation or its restoration. He drew a sharp distinction between two kinds of human impact. One he called the exploitation of nature, and the other its regulation. The former—characteristic, he thought, of his own age—was selfish, destructive, and unsustainable, yet it was the latter that involved by far the more profound and thorough transformation of nature from its existing state.[98] Fedorov expressed alarm at the evidence of environmental exploitation around him: rampant deforestation and the impoverishment of the soil by greedy and wasteful misuse, with the increase in droughts, floods, and harvest variability that seemed to have resulted.[99] But these abuses, to his mind, only intensified the problems already rife in unregulated nature that demanded human intervention. He emphatically rejected the boastful claim of nineteenth-century Western thinkers that society had achieved or was achieving a rational dominion over nature. Fedorov likened such things as laboratory science, industry, and even networks of roads and canals, the evidence usually proffered, to the works of ants compared to the power and scale of nature's processes. Human beings, he maintained, were still so far from being nature's masters that they could better be described as its slaves.[100] Deeming such bondage unacceptable, he denied at the same time that "brotherhood" with nature was a realistic or desirable ideal. He wrote as disdainfully as Lester Frank Ward in America of his educated compatriots'

[97] Ibid., I: 40, 126, 393; II: 190–191, 239; III: 382–383.

[98] Ibid., I: 72, 392–394; III: 331–332; IV: 160.

[99] Ibid., I: 38, 76–77, 196–197, 392; II: 284.

[100] Ibid., I: 393–394; II: 282–283.

worship of nature. "Is that," he asked, "a blessed enlightenment—such adoration of a blind power?"[101] His answer to the damage done to nature by heedless exploitation was not to restore it but reorganize it rationally and completely, beginning with the weather and climate. Christian scripture and tradition, he thought, expressed a deeper wisdom than any possessed by the nineteenth-century science that scorned them. In particular, the figure of the Biblical prophet Elijah implied the potential subjection of the elements to the purposes of rational beings.[102]

Fedorov grounded his ambitious proposals for controlling the atmosphere in an even more ambitious and comprehensive ethical philosophy and theology. Human death, to his mind, was the principal evil of existence, a truth he saw expressed in the Christian message of resurrection. But he interpreted that message as a call for human action to bring the dead back to life, not in the usual way as a proffered gift of miraculous divine favor. Permanent resurrection, though, could only be secured if the physical environment were brought under control and carefully regulated. Hunger, in Fedorov's view, was at bottom the one real enemy of humankind, as death's one root cause (whether directly or indirectly through warfare and disease). It could, he argued, arise from both natural and social sources. Purely social remedies that left its natural sources unaddressed were inadequate to conquer it, all the more as both overpopulation and the overexploitation and impoverishment of the planet's productive resources were already pressing against nature's limits. Only a reorganization and regulation of the physical basis of food production could permanently conquer hunger and, therefore, death itself.[103]

In this project, the weather and climate of the terrestrial sphere were the first elements that needed to be subjugated, though by no means the last. The inability of the earth, however carefully regulated, to sustain the lives of all of the resurrected dead pointed to the imperative of colonizing other planets. "A secure existence is impossible," Fedorov wrote, "so long as the earth remains isolated from other worlds. … On every planet the means of life are limited, not infinite, however large." The finiteness of

[101] Ibid., I: 76 (quotation), 106.

[102] Ibid., II: 246–249, 267; IV: 248–249.

[103] Ibid., I: 249; II: 257–259, 299–300.

the sun's energy, in the longer term, also warned humankind that it must eventually bring all cosmic as well as all terrestrial forces under regulation to assure its survival.[104] For Fedorov, regulation (for which he sometimes used the synonym "*upravlenie*," or "governance") meant creating order from nature's disorder, rational harmony out of chaos.[105]

Fedorov did not suppose that the technology for weather and climate control (let alone for human resurrection) already existed and need only be applied, but he thought that it could be developed from some promising suggestions made by earlier thinkers. One of them was V.N. Karazin (1773–1842), a scientist and founder of the University of Kharkov in the Ukraine, whom Fedorov honored as one of his great intellectual heroes.[106] Karazin had appealed (unsuccessfully) for assistance from the Russian imperial government to experiment with the application of electricity to the problems of the weather. He had planned to launch lightning rods on balloons aloft into the atmosphere as collectors of energy, and had hoped that such a means of tapping the electricity of the skies could eventually be used for weather control, particularly to cause or to prevent rainfall. Fedorov coined a term to distinguish Karazin from a conventional scientific student of the weather. The latter, he said, practiced the science of "meteorology"; Karazin had pioneered that of "meteorurgy." Meteorology's ultimate goal was the accurate forecasting of the weather, meteorurgy's its successful modification, to prevent harvest failures and other calamities.[107] The ability to make it rain, such as Karazin had proposed developing, could have forestalled the drought of 1891. The same power, Fedorov presumed, could be applied to prevent excessive rainfall that caused floods and the waterlogging of crops. Seeing electricity as a key substance underlying not only the processes of the weather, but also those of the solid earth and of the wider cosmos, Fedorov took off from Karazin's projects to speculate on the possible effects on the globe of the network of telegraphic wires that human enterprise had already begun to construct. It might well, he believed, act upon the earth's magnetic

---

[104] Ibid., I: 196–197, 202, 249–250 (quotation), 255–256, 303–303.

[105] See, e.g., ibid., II: 268, 269.

[106] Young, *The Russian Cosmists*, 12–14; Fedorov, *Sobranie sochinenii*, II: 271–272; III: 125–143.

[107] Fedorov, *Sobranie sochinenii*, II: 260–264.

field and atmosphere in a way that would make their regulation possible, ensuring the weather needed for an adequate harvest in all parts of the world simultaneously.[108]

An incident that occurred in the summer of 1891 called another possible means of weather control to Fedorov's attention. In his 1871 book *War and the Weather*, the American engineer Edward Powers accumulated a mass of circumstantial evidence linking artillery fire in the American Civil War and earlier conflicts to the subsequent occurrence of rainstorms. Powers suggested that, if a causal connection really existed, rain might be generated artificially by the deliberate production of the same kinds of loud noises. In 1890 and again in 1891, in response to severe droughts in the American Plains states, the US Congress appropriated funds for experimental tests of the Powers method. Those undertaken by a government expedition in Texas in the summer of 1891 included the cannonading of the skies, the explosion of blasting powder on the ground, and the detonation in the air of balloons containing a mixture of oxygen and hydrogen.[109]

When some of the expedition's early trial explosions were followed by clouds and rain, a number of newspapers prematurely claimed success for the method. This optimistic view reached Russia in August in an article in the newspaper *Russkie vedomosti*. Fedorov read it and would later recall that it was this article, in conjunction with the Russian famine then taking hold, that had inspired his thinking about how the military power that the nations of the world already possessed might be made a force for regulating nature.[110] Although the American experiments ended in failure, he would still speak of them occasionally as a "dazzling success" or invoke the Powers method as a promising and not yet disproven one that merited further tests. He praised the USA, by contrast with Russia and Western Europe, for having had the public spirit to fund such an important effort towards weather control.[111] He criticized it sharply, however, for the ignobly selfish and materialistic spirit in which it had pursued the

[108] Ibid., I: 76–77; II: 245–246.

[109] Spence, Clark C., "The Dyrenforth Rainmaking Experiments: A Government Venture in 'Pluviculture,'" *Arizona and the West*, 3, # 3 (1961), 205–232.

[110] Fedorov, *Sobranie sochinenii* IV: 312; V: 270.

[111] Ibid., I: 38–39, 75; II: 269–270; IV: 244.

matter and for thinking of it only in piecemeal terms rather than as a means for bringing the processes of the entire atmosphere under systematic regulation.[112] New droughts and famines in India and Russia in the late 1890s, occurring simultaneously with damaging floods elsewhere, further underlined for Fedorov the disordered state of the natural world and the need to regulate it, while he found suggestive supporting evidence for the Powers method in the war between Greece and Turkey in those years and its coincidence with unusual weather in the region.[113] Towards the end of his life, he was much taken with some new and revived projects in several European countries for preventing hailstorms by means ranging from the firing of cannons to the ringing of bells, seeing them as additional techniques whereby humans might manage the elements.[114]

By Fedorov's own account, his ideas about weather control put him at odds with members of the Russian Orthodox hierarchy. He was particularly nettled by a sermon by the archbishop of Kharkov in 1892 that rebuked the impiety and presumptuousness of attempts by human beings to make rain. The criticism itself, Fedorov maintained, was foreign to Christianity, amounting as it did to a pagan and idolatrous worship of the atmosphere, a part of the natural world.[115] He denied that there could be anything objectionable about artificially produced rain, for there was nothing more ethically objectionable than for rational creatures to bow submissively to blind, irrational forces if they could control them.[116] Many Christians, he added, accepted the justice of some forms of warfare, those that protected the weak from abuse by the strong; turning the armed forces and their weaponry into agents of weather regulation was still less objectionable, for it protected all human beings in their weakness from abuse by nature's blind powers.[117] And no one, Fedorov pointed out, had ever supposed that the Christian prayer for daily bread was meant to

[112] Ibid., I: 39; II: 273; IV: 18–19, 180.
[113] Ibid., II: 267–268.
[114] Ibid., II: 279–281; IV: 189–192, 395.
[115] Ibid., IV: 19; V: 24.
[116] Ibid., I: 40, 61; IV: 264.
[117] Ibid., I: 38–39; IV: 178.

exclude human effort to produce it, nor was irrigation seen as impious.[118] When the Bible promised a new heaven and a new earth, he reasoned, its plain meaning was that the old ones were not perfect and were subject to improvement.[119] In the same vein, he cited the familiar Biblical command to subdue the earth as justification for the task of environmental regulation. "God, having created man, commanded him to rule the earth and all that is on it. Why is it sinful to direct a cloud away from a place where heavy rain would cause harm to a place where it would be beneficial, why is it impious, why is it an offense against God?"[120] He saw not chance, but a sign from above, in the coincidence of the Powers experiments in the USA with drought and famine in Russia in the same years.[121]

A number of factors, Fedorov argued, made his own country the natural leader in the project of regulating the global environment. One of them was its geographical location. Continental rather than maritime, it possessed a harsher climate than the oceanic realm of western Europe and North America and suffered from more frequent and severe droughts. As a result, its people suffered more from an unregulated global climate and were more conscious of its evils.[122] Russia also possessed economic and political advantages for the task. What was widely regarded as its economic backwardness meant rather, to Fedorov, that it had not yet been so thoroughly infected with the evils of a capitalist economy as more "developed" countries had been. Societies more and more governed by the motive of profit had perverted science from a tool of religion, properly understood, and of the sacred obligations of human beings as conscious and rational creatures into one of capitalistic industry and commerce, dominated by city dwellers at the expense of the countryside. In such a world, even the development of accurate weather forecasting would only be turned to the profit of urban merchants.[123] The partial survival of communal peasant institutions in Russia, Fedorov believed, offered a

[118] Ibid., I: 39; II: 270–271.

[119] Ibid., IV: 181–182.

[120] Ibid., I: 39 (quotation); IV: 372.

[121] Ibid., IV: 178.

[122] Ibid., I: 76–77, 199–200.

[123] Ibid., I: 110, 129; II: 36, 284.

better foundation for the beginnings of the common task than where selfish individualism that had developed according to the mere cash nexus.[124]

Politically, too, Russia had institutions more promising than the futile and divisive parliamentary democracies of the West for a project requiring unity and devotion to a common ideal. Specifically, Fedorov's model state was one of "autocracy" (*samoderzhavie*), but the word is a tricky one. In common usage, it has come to mean (and already meant by Fedorov's time) despotic government by a single ruler. Etymologically, however, it denotes self-government (the Russian and English words are perfect cognates in this regard). Fedorov favorably contrasted autocracy in this sense of the term to two other forms of government. True autocracy, he wrote, was not despotism, which openly placed the subject under the arbitrary yoke of the ruler's will. Neither was it the same as the liberal constitutionalism of the capitalistic West, which left human beings equally unfree, though in a more hidden way. Extorting obedience to their commands through the fear of punishment, both rested illegitimately upon compulsion or force. Russian political culture had retained the name and the idea of a different kind of association that was aptly termed autocracy. The leader-autocrat did not command by force or fear, but united, coordinated, and directed the activities of the human beings who had recognized their duty of engaging in the common task.[125] Those human beings possessed freedom in the only sense that Fedorov thought the word meaningful, that of voluntarily acknowledging their obligations and choosing to fulfill them in concert with others. Mankind, he wrote, "will direct nature once there are no longer divisions within mankind, once willfulness gives way to good will."[126] An enlightened Russian tsar was the natural leader of all humankind in its projects of regulation and resurrection. Fedorov seems to have persuaded himself that the imperial Russia of his own day was an autocracy in his sense of the word and that, some malcontents from the urban intelligentsia aside, the vast majority of the Russian people gave it their free and voluntary obedience.

---

[124] Ibid., I: 253.

[125] Ibid., I: 403–409; II: 13–38; III: 387–389.

[126] Ibid., I: 393 (quotation), II: 77–78.

For Fedorov, the global scale on which effective meteorological regulation was alone possible required the unification of humankind under such leadership. The change would also entail the return of urban populations to the land and the liberation of humanity from what he called "the yoke of capitalism," in which he also saw the roots and sources of modern militarism. By relieving the pressure of population on increasingly scarce resources in the near term, his project would make the abandonment of warfare among people more feasible. Moreover, by bringing all nations into mutual interdependence in a global system of weather regulation, it would make future wars so disruptive as to become unthinkable.[127] The logic of the common task entailed unity, not only across space, but across time, between the successive generations of the human race. Fedorov saw the world's present occupants as responsible first and foremost not to the future but to the past, to their dead forbears who had given them life. Reprehensibly, they had forgotten their duty and had come to view the goal of living as mere enjoyment. Neglecting their duty to work towards the resurrection of the dead, they were themselves pushed aside and discarded by the next generation. The projects of resurrection and regulation fulfilled, in Fedorov's scheme of things, the duties of sons to their fathers.

In short, Fedorov did not regard global (and indeed cosmic) environmental engineering as a merely technical project; it was necessarily a social and political one as well. Far from seeing it as ethically problematic, moreover, he thought it ethically imperative upon humankind, but feasible only under a transformed regime of the political, economic, and social arrangements of the human race. Throughout his writings, he expressed a vehement scorn for "progress" and the forms in which the nineteenth century had idolized it. But far from renouncing the hope of progress, he merely rejected the usual understanding of what it involved: the expansion of a shallow material prosperity and of individual caprice. Global human union in his common task, to his mind, alone represented a worthwhile goal toward which genuine progress could be made.

A younger Russian philosopher accurately described Fedorov as a hater of capitalism and of unfettered individualism, a collectivist, and a

[127] Ibid., I: 39–40, 77; II: 34, 273–275, 322–325; IV: 181.

prophet of world government and of global-scale planning.[128] It would be difficult to assemble a list of traits less characteristic of today's principal Prometheans, and particularly the advocates among them of the kind of global climatic engineering that Fedorov proposed. Even so, he was the one major environmental Promethean of his time (and it would be difficult to find a more Promethean thinker of any time) who at least on the face of things cannot simply be labeled as a political or social progressive. For all its unequalled environmental radicalism, his social thought, rooted as it was in Christian scripture and Russian Orthodox tradition, has elements better described as backward-looking, down to its particular concern for the welfare of past generations. But the matter does not end there. However Fedorov himself may have looked at things, the reception of his project suggests that others found it the opposite of conservative. It seems to have aroused no interest whatsoever within the Tsarist government or, criticism aside, the Orthodox Church. It aroused a great deal among Russian political and artistic radicals of the early twentieth century. Traces of his ideas are apparent in many of the foremost Russian writers of the first few decades of the twentieth century, including Aleksandr Blok, Maksim Gorky, Vladimir Mayakovsky, Nikolai Zabolotsky, and Boris Pasternak, all of them unsympathetic to the old political and social order in Russia; a number of artists of the early twentieth-century avant-garde; and most of the significant Russian philosophers of the same era, few of them, again, apologists for the status quo. They can also be seen in the work of V.I. Vernadsky, and particularly in Vernadsky's concept of the "noosphere" as the earth's surface envelope, a biosphere suffused and physically transformed by the work of human thought, and in Vernadsky's untroubled attitude toward its emergence.[129]

It is particularly difficult to trace the relation of Fedorov's work to the ideals and policies of the Soviet Union after 1917, when Russian writers, living under an aggressively anti-religious state, may well have been reluctant

[128] Nicolas Berdyaev, *The Origins of Russian Communism*, trans. R. M. French (London: G. Bles, 1948), 93.

[129] For Fedorov's influence on these and other avant-garde figures, see Young, *The Russian Cosmists*; Irene Masing-Delic, *Abolishing Death: A Salvation Myth of Russian Twentieth-Century Literature* (Stanford, CA: Stanford University Press, 1992); and John E. Bowlt and Olga Matich, *The Russian Avant-Garde and Cultural Experiment* (Stanford, CA: Stanford University Press, 1996).

to acknowledge openly any indebtedness to an overtly (though quite unorthodox) Christian thinker. Yet, as the historian Dmitri Shlapentokh has documented, not merely was there much common ground, in the mutual goal of a Promethean conquest of nature by a collectivist society, but channels of possible influence can be traced from Fedorov and his disciples to such individual Bolshevik leaders as Trotsky and Stalin.[130] The affinities may also be the result not of direct contact but of shared elements in the Russian tradition. The conquest of nature had been an important theme in nineteenth-century Russian radical thought, which more generally had shown far more affinity for the physical and applied sciences of nature than for the ecological ones.[131] The affinities may also reflect similar elements in Marxism that overlapped with Fedorov's ideas. In any case, as Shlapentokh noted, the common ground may have made it easier for Fedorovians to accept the Bolshevik regime, which preached the conquest of the natural world, the climate included, and which might easily have been justified in Fedorovian terms, as an autocracy (in the more common sense of the word) directing people in the way they would have voluntarily chosen to go if they had been sufficiently enlightened.

Leon Trotsky predicted in the 1920s that, under socialism:

> nature will become more 'artificial.' The present distribution of mountains and rivers, of fields, of meadows, of steppes, of forests, and of seashores, cannot be considered final. Man has already made changes in the map of nature that are not few nor insignificant. But they are mere pupils' practice in comparison with what is coming. Faith merely promises to move mountains; but technology, which takes nothing "on faith," is actually able to cut down mountains and move them. … in the future, this will be done on an immeasurably larger scale, according to a general industrial and artistic plan. Man will occupy himself with re-registering mountains and rivers,

---

130 Dmitry Shlapentokh, "Bolshevism as a Fedorovian Regime," *Cahiers du monde russe* 37, #4 (1996), 429–465.

131 Masing-Delic, *Abolishing Death*, 66–72; Catherine Evtuhov, *The Cross and the Sickle: Sergei Bulgakov and the Fate of Russian Religious Philosophy* (Ithaca, NY: Cornell University Press, 1997), 68–75; Simon Karlinsky, "Nabokov and Chekhov: The Lesser Russian Tradition," in Alfred Appel, Jr. and Charles Newman, eds., *Nabokov: Criticism, Reminiscences, Translations, and Tributes* (Evanston, IL: Northwestern University Press, 1970), 7–16.

and will earnestly and repeatedly make improvements in nature. In the end, he will have rebuilt the earth, if not in his own image, at least according to his own taste.[132]

Vaulting Promethean ambitions for the conquest of the earth were a frequent note struck by Soviet writers in the 1920s.[133] They remained prominent, though in a different key, under the doctrines of Socialist Realism imposed on writers and artists by Trotsky's victorious adversary, Stalin. The often delirious earlier visions of human mastery gave way to more prosaic accounts of the conquest of nature and the transformation of the landscape by the loyal representatives of the communist state, who irrigated deserts, drained wetlands, expanded civilization northward into the Arctic, and tamed and harnessed rivers by dams and hydroelectric power stations.[134] Stalin and his successors into the 1960s pursued all of these undertakings in fact as well as in fiction, to create showpieces of "socialist construction." Their apparent success, illustrating the technological and organizational might of Marxist-Leninist societies as compared with the feebleness of Western capitalist ones, won the plaudits of such sympathetic foreign visitors as J.D. Bernal. The Soviet regime shared

[132] Leon Trotsky, *Literature and Revolution*, trans. Rose Strunsky (New York, NY: International Publishers, 1925), 251. On Trotsky's Prometheanism, see also Irene Masing-Delic, "Boris Pilniak's *The Volga Falls to the Caspian Sea* as Trotskyite Sophiology," *Slavic and East European Journal* 52, #3 (2008), 414–438.

[133] Stites, *Revolutionary Dreams*, ch. 8; Anthony J. Vanchu, "Technology as Esoteric Cosmology in Early Soviet Literature," in Bernice Glatzer Rosenthal, ed., *The Occult in Russian and Soviet Culture* (Ithaca, NY: Cornell University Press, 1997), 203–222; Irene Masing-Delic, "Zabolotsky's The Triumph of Agriculture: Satire or Utopia?," *Russian Review* 42, #4 (1983), 360–376.

[134] On these works and the importance of the theme of conquering nature in Stalinist Socialist Realism more generally, see Mary A. Nicholas, *Writers at Work: Russian Production Novels and the Construction of Soviet Culture* (Lewisburg, PA: Bucknell University Press, 2010); Frank Westerman, *Engineers of the Soul: The Grandiose Propaganda of Stalin's Russia*, trans. Sam Garrett (New York, NY: The Overlook Press, 2011); Marc Slonim, *Soviet Russian Literature: Writers and Problems, 1917–1977*, revised edition (New York, NY: Oxford University Press, 1977), 241–247; and William B. Husband, "'Correcting Nature's Mistakes': Transforming the Environment and Soviet Children's Literature, 1928–1941," *Environmental History* 11, #2 (2006), 300–318. On the same theme in Soviet cinema, see Evgeny Dobrenko, *Political Economy of Socialist Realism*, trans. Jesse M. Savage (New Haven, CT: Yale University Press, 2007), 312–327, and on its interaction with the traditional esthetics of nature in Soviet landscape painting, Mark Bassin, "'I Object to Rain that is Cheerless': Landscape Art and the Stalinist Aesthetic Imagination," *Ecumene* 7, #3 (2000), 313–336.

Bernal's enthusiasm for the use of nuclear power to carve new landscapes and to master climate at regional and global scales.[135]

Such a Promethean relation to geographical nature ran parallel to the anti-Mendelian biological theories of T.D. Lysenko, which became Soviet dogma in the 1940s, with disastrous effects on Russian agronomy and genetics. Lysenko himself was involved in the development of the so-called Stalin Plan for the Transformation of Nature, first decreed in 1948 and further elaborated over the next several years.[136] The Plan proposed an enormous effort of afforestation, pond and reservoir construction, and irrigation, on paper much what Ignatius Donnelly had proposed for the western USA, to conquer the hostile forces of wind and drought and extend the secure area of cultivation on the semiarid steppes of the south.[137] Relatively muted in the area of forestry by some countervening

---

[135] Paul R. Josephson, *Red Atom: Russia's Nuclear Power Program from Stalin to Today* (New York, NY: W. H. Freeman and Company, 2000), 245–250; Paul R. Josephson, "'Projects of the Century' in Soviet History: Large-Scale Technologies from Lenin to Gorbachev," *Technology and Culture* 36, #3 (1995), 519–559; James Rodger Fleming, *Fixing the Sky: The Checkered History of Weather and Climate Control* (New York, NY: Columbia University Press, 2010), 198–200.

[136] Valery N. Soyfer, *Lysenko and the Tragedy of Soviet Science*, trans. Leo Gruliow and Rebecca Gruliow (New Brunswick, NJ; Rutgers University Press, 1994), 205–209.

[137] Douglas Weiner, *A Little Corner of Freedom: Russian Nature Protection from Stalin to Gorbachev* (Berkeley, CA: University of California Press, 1999), 88–93; Denis J. B. Shaw, "Mastering Nature Through Science: Soviet Geographers and the Great Stalin Plan for the Transformation of Nature, 1948–1953," *Slavonic and East European Review* 93, #1 (2015), 120–146. Stephen Brain, *Song of the Forest: Russian Forestry and Stalinist Environmentalism* (Pittsburgh, PA: University of Pittsburgh Press, 2011), has made large claims for the role of "Stalinist environmentalism" in Soviet forest policy of the 1930s and 1940s. He has rightly called attention to Stalin's reservation of substantial areas of Russian forest from cutting or other intensive use, and he has presented the Stalin Plan of 1948, at least for some of its architects, as one not of Promethean transformation, but of environmental restoration, based on pre-Soviet theories about the originally wooded character of much of the steppe and its damaging clearance by settlement. The argument, however, requires only modest qualification of the accepted view of Soviet Prometheanism. Brain did not demonstrate that the Soviet leaders who launched the plan accepted the theory that the steppes had originally been forested, and he himself throughout described the goal of the Plan as "afforestation," not as "reforestation," just as the plan was avowedly one of "transformation" and not of "restoration." He noted the involvement in its development and implementation of a large cadre of Prometheans, whose views are apparent in the inclusion of irrigation and pond-construction projects—certainly not ecologically restorationist by any standards—even in the original plan, whose elaboration in 1950, extending its area well to the east, added further major projects of river control and of canal construction. Even if one were to waive all of these points, Brain's argument applies only to the preservation or restoration of forests, whose privileged place in Russian culture may have made them an exception (see, e.g., Bassin, "I Object to Rain that is Cheerless,"), and not to the rest of the environment,

factors, Stalinist and post-Stalinist Prometheanism ran unchecked in other areas. As an overarching earthview, it received official codification in the entry on "Geographical Environment" in the *Great Soviet Encyclopedia* of 1952, which distinguished sharply between the environmental relations of capitalist and communist countries. Both kinds of societies altered nature; their difference did not consist in that, but in the way in which they did so. Anarchic competition for profit under capitalism devastated and impoverished the material basis of human well-being. Rational planning under socialism transformed nature in the service of society. The American Dust Bowl epitomized the former; the Stalin Plan, with its components of afforestation, large-scale irrigation, and great hydroelectric stations, the latter.[138]

The socialist conquest of nature found a forceful and prestigious spokesman in the writer Maxim Gorky (1868–1936). He became a kind of literary laureate of the idea, as well as of the Soviet regime in general, when he returned permanently to Russia in the early 1930s. In Douglas Weiner's words, Gorky's "commitment to a total transformation of human and nonhuman nature—by human reason—was at the core of his intellectual life and *opus*."[139] His vehement atheism led Gorky to advocate "God-building" by humankind and the assumption of the Creator's role over the material world. Gorky himself summed up his attitude in approvingly quoting an early Bolshevik stalwart: "Let's declare war on nature." The violence and the comprehensiveness of his rhetoric can startle a reader today. As Weiner wrote: "Swamps, predators, snakes, Arctic ice, hurricanes, earthquakes, unproductive deserts,

---

which was protected by no such special status—and whose partial protection would also constitute the kind of picking and choosing among elements of nature that is already more Promethean than environmentalist. The term "environmentalist," if it means anything, means a generalized earthview and not actions restricted to a single realm of the environment and combined with Promethean projects and rhetoric in all others.

[138] "Geograficheskaia sreda," in V. A. Vvedenskii, ed., *Bol'shaia sovetskaia entsiklopediia*, vol. 10 (Moscow: Gosudarstvennoe nauchnoe izdatel'stvo, 1952), 452–453.

[139] Douglas R. Weiner, "Man of Plastic: Gor'kii's Visions of Humans in Nature," *The Soviet and Post-Soviet Review* 22, #1 (1995), 65–88 (quotation from 65). Another excellent discussion of Gorky's uncompromising hostility to the natural environment is Dobrenko, *Political Economy of Socialist Realism*, 76–83. See also Masing-Delic, *Abolishing Death*, 72–75.

drought, sleepy forests and parasites would all be eliminated." Like Comte a century earlier, Gorky deplored the way in which poets had continued to glorify nature, despicably praising an evil tyrant rather than the heroic and enlightened human beings who sought to remake it and to banish its evils. But only under socialist planning, he added, not under capitalism, could they do so with success. The superiority of the former was evident in nothing so much as its far greater ability to rebuild the earth.[140] His fusion of Prometheanism and Soviet socialism led him into the project that has done more to blacken his reputation than any other, coordinating the work of a number of writers into an anthology praising the construction by the forced labor of the White Sea-Baltic Canal in the mid-1930s. Gorky and his associates presented the work as simultaneously a great remaking of geography and a means for the education and redemption by useful work of the mostly political prisoners assigned to the task.[141]

A history of censorship has long made nature in Russian literature as much a code for comment on the sociopolitical world as a realm in its own right, and it is no accident that it became a symbol for two notable skeptics of the Soviet regime. In one of the most notable dystopian novels ever published, Evgeny Zamyatin's *My* (*We*), a "green wall" separates a city of an almost Gastevian degree of regimentation, physical and social, from wild nature, distrusted and feared.[142] Andrei Platonov (1899–1951) began the 1920s attempting as a reclamation engineer to put into practice a combination of Fedorovian and Bolshevik enthusiasm; by later in the decade and thereafter, his work began to stress both

---

[140] M. Gorky, "O bor'be s prirodoi," in F. N. Petrov, ed., *Gor'kii i nauka: Stat'i, rechi, pis'ma, vospominaniia* (Moscow: Nauka, 1964), 186–198 (quotation from 186); "O biblioteke poeta," in ibid., 176–18; "Zasukha budet unichtozhena," in ibid., 150–152; "O prave na pogode," in ibid., 236–240; Weiner, "Man of Plastic," 70.

[141] Ibid., 73–74; Dobrenko, *Political Economy*, 105–124; Dariusz Tolczyk, *See No Evil: Literary Cover-Ups and Discoveries of the Soviet Camp Experience* (New Haven, CT: Yale University Press, 1999), 150–180; Westerman, *Engineers of the Soul.*

[142] Evgeny Zamyatin, *We*, trans. Natasha Randall (New York: Modern Library, 2006); see Patricia Carden, "Utopia and Anti-Utopia: Aleksei Gastev and Evgeny Zamyatin," *Russian Review* 46, #1 (1987), 1–18.

the recalcitrance of nature (canals that would not fill, deserts that would not blossom) and the human costs of impatient and imperious planning from above.[143] Where utopia is Promethean, dystopia becomes environmentalist and organic.

[143] The literature on Platonov, including investigations of his relation to Fedorov, is quite large, but see, e.g., Thomas Seifrid, *Andrei Platonov: Uncertainties of Spirit* (Cambridge: Cambridge University Press, 1992) and *A Companion to Andrei Platonov's The Foundation Pit* (Boston, MA: Academic Studies Press, 2009) and Mieka Erley, "'The Dialectics of Nature in Kara-Kum': Andrei Platonov's *Dzhan* as the Environmental History of a Future Utopia," *Slavic Review* 73, #3 (2014), 727–750. On the influence on his earlier work of Gastev and especially Bogdanov, see Seifrid, *A Companion*, 38–42, 145–147 and Elena Tolstaia-Segal, "Ideologicheskie konteksty Platonova," *Russian Literature* 9, #3 (1981), 238–241.

5

Conclusion: The Politics of Prometheanism Revisited

I began by asking whether there were notable advocates of the earthview of environmental Prometheanism on the left-hand side of the political spectrum in the period and countries examined. One point the book has established is that there were. Left-wing Prometheans, who seem not to have existed in any sizeable numbers or prominence in recent times, did so in the past: not merely cranks on the margins of intellectual life in their day, but thinkers of considerable stature and influence who, moreover, presented their worldviews and earthviews as closely and necessarily connected. They subscribed, as a rule, to both of the specific sub-themes, of progress and of egalitarianism, that most reliably characterize the left. For the most part, they valued tradition and continuity in human society much less than they did the possibilities of change for the better. Most were notably egalitarian by the standards of their time and place. In these respects, they differed dramatically from the most visible modern Prometheans, who have tended more to be found on the political right, to place a higher value on historical tradition, and to be more tolerant of social inequality than most environmentalists, their antitheses in earthviews. The progressive Prometheanism that this book has

W.B. Meyer, *The Progressive Environmental Prometheans*,
DOI 10.1007/978-3-319-29263-2_5

documented represents a once-considerable and now almost vanished environmental discourse. What one might infer from modern politics, that Prometheanism and leftism are at some level basically antagonistic, is plainly not the case. The question that then arises is why they have parted ways.

A more tentative finding, one that adds to the cogency of that question, is that progressive or left-wing Prometheanism not only existed in the past, but was as much its dominant form of anti-environmentalism as right-wing Prometheanism later became. For in searching for the most notable and articulate past champions of a Promethean earthview, I found none on the political right comparable in stature or influence to those I have described on the left. N.F. Fedorov, many of whose views seem more characteristic of the right than the left, nonetheless on closer examination confirms the rule more than he contradicts it, for whatever his own politics may have been, Russian progressives found his ideas far more attractive than conservatives did. Prometheanism's prevalent ideological affiliations seem to have undergone a reversal between the era examined in this book and the present. The political worldview it was associated with then is the opposite of the one it keeps company with now. If true, the finding weakens the hope that we can discover some deep-seated and timeless logic to today's pattern, and it further whets one's curiosity as to why such a reversal should have happened.

So does one more aspect of the history the book has recounted. Another characteristic of modern environmental Prometheans, anti-statism or at least what goes by that name, was not at all characteristic of their predecessors. Most of the past Prometheans examined here—and this time Fedorov among them—vocally advocated a large and, indeed, expanded role for a central, sometimes even a world government. They agreed with Lester Frank Ward and H.G. Wells that the interests of human beings in their dealings both with one another and with nature suffered not from too strong a state but too weak a one. Such anarchists as Reclus and Kropotkin differed on this point, but they did not advocate anything like a present-day market libertarianism. Putting no trust in the centralized nation-state, they equally scorned the pure individualism of the capitalist market and placed their hopes in the collective local commune, which modern libertarians would see as more oppressively state-like than not.

The other earlier Prometheans, uniting their faith in the state with a program of mastering and reshaping the earth's surface, could hardly differ more from their right-wing successors.

Skeptics might challenge the evidence I offer for the scarcity of right-wing or laissez-faire Prometheanism in the earlier period as inadequate. It is—though unavoidably if the claim is correct—mostly negative evidence, the least compelling kind. I tried to identify and discuss the most prominent and articulate Promethean thinkers of the era, and I found them on the left. I failed to identify any large number of equally distinguished and thoughtful conservative enthusiasts for the engineering of the earth's surface. But perhaps the failure was my own. The history of Promethean ideas that I tried to survey is, as I observed in the introduction, poorly charted terrain indeed, and the mists of ignorance or preconception may have hidden even some of its highest peaks from my view. Can I offer any other grounds for the conclusion that Prometheanism was not as common among conservative as among progressive thinkers? Was it not simply a part of the entire climate of western opinion of the time, as John Dryzek has argued, a set of presuppositions that more or less everyone before about 1960 held, whatever their other differences? It is true that most of the thinkers examined here presented their environmental Prometheanism as necessarily related to their progressivism. But modern right-wing Prometheans equally present theirs as fully consistent with their conservatism. It is entirely possible that those of 1790–1960, if there were any, might have done the same.

Nothing is more likely than that I have missed some important exponents of environmental Prometheanism in the four countries in question. It is much less likely than that I have missed so many, and in so unevenly distributed a way, that a full accounting would show right-wing Prometheans of the era in question in the countries examined to be as prevalent as left-wing ones or more so. Even so, my conclusions invite a further test. Among the right-wing figures in each country who were comparable in importance to the most historically and politically important of the progressive Prometheans I have examined, what environmental views prevailed?

I am competent to answer in detail only for the USA. Barlow and the northern Jeffersonians, the American Fourierists, Greeley, Donnelly,

Howells, Ward, Gilman, the conservationists and the Progressives of the early and mid-twentieth century, Dewey, Beard, Veblen, and the institutional economists are all notable figures in the history of American progressive or left-wing political thought (and most of them also that of opposition to economic laissez-faire). We can look for comparison at the equally major figures in pre-1960 American conservatism, the ones discussed in surveys by Russell Kirk (*The Conservative Mind*), Allen Guttmann (*The Conservative Tradition in America*), Clinton Rossiter (*Conservatism in America*), and Patrick Allitt (*The Conservatives*).[1] Other than the antebellum Southern agricultural reformer Edmund Ruffin (himself only a minor figure in the accounts of the two who mention him), we find no one who, to my knowledge, was a similarly outspoken proponent of environmental Prometheanism.[2] Instead, within their central cast of characters we find some notable skeptics of the Promethean project: of the project itself, that is, not merely of the way in which a particular kind of society of which they disapproved was mishandling it. Joel Barlow's expansive and utterly untroubled confidence about human-induced change contrasts strongly with the views of his contemporary and onetime friend, the unbending Federalist conservative Timothy Dwight, who was deeply ambivalent about the progress of human improvement in the landscape and "decidedly anxious about the durability of environmental resources" unless proper curbs were put on their exploitation.[3]

[1] Russell Kirk, *The Conservative Mind, from Burke to Santayana* (Chicago, IL: Henry Regnery Company, 1953); Allen Guttmann, *The Conservative Tradition in America* (New York, NY: Oxford University Press, 1967); Clinton Rossiter, *Conservatism in America*, revised edition (Cambridge, MA: Harvard University Press, 1982); Patrick Allitt, *The Conservatives: Ideas and Personalities Throughout American History* (New Haven, CT: Yale University Press, 2009).

[2] On Ruffin's Prometheanism, see Jack Temple Kirby, *Poquosin: A Study of Rural Landscape & Society* (Chapel Hill, NC: University of North Carolina Press, 1995), 85. Ruffin is mentioned in passing, as an antebellum apologist for slavery, by Rossiter, *Conservatism in America*, 126 and Allitt, *The Conservatives*, 40, 87.

[3] Jane Kamensky, "In These Contrasted Climes, How Chang'd the Scene: Progress, Declension, and Balance in the Landscapes of Timothy Dwight," *New England Quarterly* 63, #1 (1990), 80–108; Thomas Hallock, *From the Fallen Tree: Frontier Narratives, Environmental Politics, and the Roots of a National Past* (Chapel Hill, NC: University of North Carolina Press, 2003), 186–195; Lloyd Willis, *Environmental Evasion: The Literary, Critical, and Cultural Politics of Nature's Nation* (Albany, NY: SUNY Press, 2011), 45–47 (quotation from 45). Dwight also expected the deforestation that Barlow thought would produce a milder American climate to create instead a more severe one: Timothy Dwight, *Travels in New England and New York*, ed. Barbara Miller Solomon, vol. 1 (Cambridge, MA: Belknap Press of Harvard University Press, 1969), 40–41. Dwight is discussed

The novelist and conservative political observer James Fenimore Cooper made his most memorable and eloquent fictional character a vocal critic of the transformation of nature.[4] The Harvard professor of fine arts Charles Eliot Norton played a leading role in the campaign to save the beauties of Niagara Falls from industrial and commercial development; the Bostonian historian Francis Parkman was an early advocate of wilderness preservation.[5] The historian Henry Adams, at least in some moods, invoked the physics of energy and the depletion of mineral deposits to warn that the demands of an expanding human civilization were running up against the limits of a finite terrestrial resource base.[6] Madison Grant combined conservative politics and theories of Nordic racial superiority with environmental preservationism.[7] The poet-critic T.S. Eliot wrote that the "exploitation of the earth" in modern times threatened "the permanent conditions upon which God allows us to live on this planet," and penned the lines: "Schemes of human greatness thoroughly discredited/ Binding the earth and the water to your service/Exploiting the seas and

---

by Rossiter, *American Conservatism*, 116, 125 and mentioned by Guttman, *The Conservative Tradition*, 81 and Allitt, *The Conservatives*, 18–19.

[4] Willis, *Environmental Evasion*, Chap. 2; Annette Kolodny, *The Lay of the Land: Metaphor as Experience and History in American Life and Letters* (Chapel Hill, NC: University of North Carolina Press, 1975), 89–115; Nelson Van Valen, "James Fenimore Cooper and the Conservation Schism," *New York History* 62, #3 (1981), 289–306. All four books discuss Cooper as an important American conservative.

[5] On Norton's preservationist activities, see Kermit Vanderbilt, *Charles Eliot Norton: Apostle of Culture in a Democracy* (Cambridge, MA: Harvard University Press, 1959), 188–190 and James Turner, *The Liberal Education of Charles Eliot Norton* (Baltimore, MD: Johns Hopkins University Press, 1999), 282, 290, 297, 299, 309; on his politics, Rossiter, *Conservatism in America*, 159; Allitt, *The Conservatives*, 107–109. On Parkman, Nash, *Wilderness and the American Mind*, 98–100; Wilbur R. Jacobs, "Francis Parkman: Naturalist-Environmental Savant," *Pacific Historical Review* 61, #3 (1992), 343–347; Guttmann, *The Conservative Tradition*, 123; and Allitt, *The Conservatives*, 68–69, 71–72.

[6] Henry Adams, *A Letter to American Teachers of History* (Washington, DC: n.p. 1910), 130–136; J. C. Levenson et al., eds. *The Letters of Henry Adams* (Cambridge, MA: Belknap Press of Harvard University Press, 1988), vol. 4 (1892–1899), 722–723, 733; vol. 5 (1899–1905), 251, 444–445, 590; vol. 6 (1906–1918), 126, 506. (In other moods, Adams sometimes spoke of the likelihood that accelerating liberation of energy from nature would destroy civilization, an opposite argument but an equally anti-Promethean one.) All four histories treat Adams as a significant figure in American conservative thought, though Guttmann, *The Conservative Tradition*, differs from the other three in calling him "a conservative *manqué*" (135).

[7] Spiro, *Defending the Master Race*; Rossiter, *Conservatism in America*, 159; Allitt, *The Conservatives*, 120.

developing the mountains."[8] The Southern Agrarians of the 1930s—like Eliot, principled inegalitarians and traditionalists—dwelt in their manifesto *I'll Take My Stand* (1930) on the evils and illusions of modern society's project of conquering or remaking the natural environment.[9] Another Southern conservative, Richard Weaver, was a kindred spirit, though not a formal member of the group. His most important work, *Ideas Have Consequences* (1948), criticized prevalent attitudes toward the order of nature: "For centuries we have been told that our happiness requires an unrelenting assault upon this order … that nature is hostile to man or that her ways are offensive or slovenly, so that every step of progress is measured by how far we have altered them." Against the manipulation of nature, Weaver argued that "to meddle with small parts of a machine of whose total design and purpose we are ignorant produces evil consequences."[10] In these writers we find expressed the very arguments, as discussed in the Introduction, that seem to make conservatives and environmentalists natural allies.

A similar examination might be worth undertaking for other countries. At least to my knowledge, the environmental Prometheanism of such progressives as the Saint-Simonians, Fourier and his followers, Comte, Reclus, Berthelot, and Teilhard was not duplicated to any significant degree in the thought of the leading French conservatives, from Maistre and Bonald in the aftermath of the Revolution to Charles Maurras and the intellectual entourage of the Vichy regime in the early to mid-twentieth century. Nor did the Prometheanism characteristic of Russian radicalism seem to have much appeal among the major figures of Russian conservatism identified by Richard Pipes in 2005.[11] Nineteenth-century

[8] T. S. Eliot, *The Idea of a Christian Society* (New York: Harcourt, Brace and Company, 1940), 62–63; "Choruses from The Rock," *Collected Poems, 1909–1935* (New York: Harcourt, Brace and Company, 1936), 192. All four historians treat Eliot as a major conservative writer.

[9] Twelve Southerners, *I'll Take My Stand: The South and the Agrarian Tradition* (New York, NY: Harper & Brothers, 1930), xiv, 5–10. See Guttmann, *The Conservative Tradition*, 148–157; Allitt, *The Conservatives*, 136–141.

[10] Richard M. Weaver, *Ideas Have Consequences* (Chicago, IL: University of Chicago Press, 1948), 171–172. All but Kirk identify Weaver as a major figure.

[11] Pipes, *Russian Conservatism and Its Critics*.

England had a significant tradition of what might be called green conservatism; I am not aware that it had one of conservative Prometheanism.[12]

It would be claiming far too much, though, to say that to be progressive was necessarily to be Promethean. Certainly many progressives ignored the environmental dimension of human life altogether, and a handful of thinkers—notably Emerson and Thoreau in the USA and John Stuart Mill, William Morris, and John Ruskin in England—have a claim, though in each case a contestable one, to be considered both progressive and environmentalist.[13] There were and are also apolitical Prometheans, enthusiastic for the technologies of environmental engineering without, at least consciously or explicitly, coupling them with any program for society. A nineteenth-century example from the USA is the meteorologist James Pollard Espy (1785–1860), who in the 1840s offered the USA a comprehensive program of control over its weather. (Espy, though, was notably progressive at least in his religious views, and his Promethean program aroused the deepest misgivings in the conservative antebellum South.[14]) Later ones include the electrical engineer Nikola Tesla (1856–1943) and the technological visionary R. Buckminster Fuller (1893–1983).

The thought of some other past figures can only be confused by the classifications of environmentalist and Promethean that I have used. Theodore Roosevelt is notable both for his actions to protect untamed wilderness and for the strongly Promethean platform, consistent with

---

[12] Katey Castallano, *The Ecology of British Romantic Conservatism, 1790–1837* (Basingstoke: Palgrave Macmillan, 2013).

[13] Emerson's and even Thoreau's earthviews were not unambiguously environmentalist: see, e.g., Robert S. Corrington, "Emerson and the Agricultural Midworld," 140–152 and Douglas R. Anderson, "Wild Farming: Thoreau and Agrarian Life," 153–163, in Paul B. Thompson and Thomas C. Hilde, eds., *The Agrarian Roots of Pragmatism* (Nashville, TN: Vanderbilt University Press, 2000). Thoreau's politics were more libertarian than progressive or egalitarian. Ruskin's were even less conventionally leftist, and he was enthusiastically involved in a number of projects for reclaiming land from its natural state. Morris was egalitarian, but (as Wells observed) traditionalist in his politics. Mill expressed the hope that some of nature would be preserved (less for its own sake or the vital functions it played than to afford human beings the occasional experience of solitude and communion with it), but also, in his essay "Nature," decried the use of nature as a standard of good and praised many changes human activity had produced in the earth's surface as unquestionably improvements.

[14] William B. Meyer, *Americans and Their Weather: A History* (New York, NY: Oxford University Press, 2000), 87–88.

many of his other policies, on which he accepted the Progressive nomination for president in 1912. At the same time, he so far escapes the usual labels of political ideology as to appear as an important figure in surveys of American progressivism and American conservatism alike (he was also an enthusiast for war and winner of the Nobel Peace Prize and a "trust-buster" who became a prophet of regulated monopoly). Probably, then, he thought in other terms altogether. He is better understood, indeed, as a prophet of heroic action and the strenuous life for their own sake, of national strength rather than personal happiness, whose political good fortune it was that, while in office, he never had to call Americans as a group to any particularly strenuous action. If he possessed a definite earthview, it was of nature as a testing ground for human beings. Its tests could be met sometimes by conquering it and transforming it, sometimes by surviving a stay in its untamed reaches (whose preservation as such was important for providing such experiences), sometimes by hard mental work to untangle its mysteries, in each case with quite different consequences for nature itself. So too, politically Roosevelt was less a progressive or a conservative than a devotee of national energy, and inclined to judge matters chiefly according to that criterion. He urged progressive reform so that what he feared as a destructive radicalism should not capture the nation-state and cripple its effectiveness. But Roosevelt was an unusual figure. Classifications are neither right nor wrong in themselves, only more or less useful, and one of the determinants of their usefulness is the number of cases to which they apply. Those of conservative and progressive, environmentalist and Promethean apply poorly to Roosevelt because he was unusual. They have come into and remained in use because they work well for most thinkers. Used on the subject matter of this book, they apply well enough generally to reveal a pattern in the more distant past just as they do one in the recent past, the former the opposite of the latter.

The four countries I have examined have, of course, their differences. France has often been described as placing relatively little value on nature, perhaps from its Cartesian heritage, in the latter because of a pervasive emphasis on spiritual over material realities. A love of nature (and dislike of the city), conversely, has been a longstanding theme in the cultural history of both England and the USA, which differ greatly, on the other hand, in the high status accorded to wilderness in the latter. Russian traditions have different notably from those of the West. Political thought in the four countries has attached different specifics to the labels of left and right. It is all the more telling, then, that all have displayed broadly similar histories, until recent decades, of a progressive Prometheanism transcending national peculiarities.

Countries other than the four that I have examined might furnish useful additional tests of my conclusions. Books by two historians offer some relevant evidence from a fifth. In *The Conquest of Nature* (2006), David Blackbourn focused more narrowly on the management of water, and his central question was not the same as mine, but he did demonstrate the existence, if not necessarily the prevalence, of progressive Prometheanism in nineteenth- and early twentieth-century Germany. "Liberals and progressives," Blackbourn wrote, often "saw themselves fighting a war on two fronts: against a nature that constrained humanity, and the 'backward-looking' humans who did the same." For their part, he continued, many conservatives resisted the siren song of environmental conquest: "In mourning lost nature, they mourned for a world in which everything they treasured—ideas and social relations as much as familiar landmarks—seemed to be at risk." Those most skeptical of the benefits of human impact, Blackbourn observes, also included Friedrich Nietzsche, the nineteenth century's most incisive critic of progressivism and egalitarianism.[15] Raymond Dominick, in *The Environmental Movement in Germany*, took pains to emphasize that concern about the environment, in various forms (not all of them "environmentalist" in my sense), was not limited to the upper classes. All the same, his evidence suggests that it was disproportionately strong there, especially among the aristocracy, and suggests, too, that the main nineteenth-century German prophets of environmentalism, the matter chiefly of interest here, were conservative rather than progressive. Concern about human impact was particularly marked, Dominick wrote among nationalist defenders of the social and political status quo, and the environmental concerns he found in such progressives as Friedrich Engels and August Bebel are more Promethean

[15] David Blackbourn, *The Conquest of Nature: Water, Landscape and the Making of Modern Germany* (London: Pimlico, 2007), 175, 180, 184.

than environmentalist.[16] And though, as Blackbourn and many others have pointed out, the characterization of National Socialism as a green or environmentalist movement is vastly oversimplified, and though there were some non-Promethean elements in Soviet Marxism, a difference in tone and emphasis existed all the same. Nazism, by most reckonings an extreme right-wing movement, included much more in the way of environmentalist ideas and rhetoric (and some environmentalist action as well) than did Russian Bolshevism, by most standards an extreme left-wing movement. Maoist China, as Judith Shapiro has shown, embraced an ideology of a "war on nature," and most would situate it, too, ideologically on the far left.[17]

I have contrasted the political affinities of Prometheanism before about 1960 with those of the present. Much might also be learned from a careful study of its associations, and environmentalism's too, in the intervening period. Jean Jacob has traced the evolution in postwar France of "political ecology," or of an environmentalism that connects its earth-view with a clear political stance. In its earliest days, by his account, it drew at least as much support and as many spokesmen from the right as from the left, but by the 1990s, it was unambiguously left-wing in its orientation.[18] The historian Brian Allen Drake has examined the careers of several prominent postwar Americans who combined a right-wing (chiefly libertarian) politics with environmentalism. (Most of them, it is perhaps relevant to note, were from the West, where government and

---

[16] Raymond Dominick, *The Environmental Movement in Germany: Prophets and Pioneers, 1871–1971* (Bloomington, IN: Indiana University Press, 1992), 22–25, 62–63. For similar evidence, see also Colin Riordan, "Green Ideas in Germany: A Historical Survey," in Colin Riordan, ed., *Green Thought in German Culture: Historical and Contemporary Perspectives* (Cardiff: University of Wales Press, 1997), 3–41. Carl Mitcham, *Thinking Through Technology: The Path Between Technology and Philosophy* (Chicago, IL: University of Chicago Press, 1994), 20–24 also discusses the work of Ernst Kapp (1808–1896), German political radical and Promethean geographer. Even in the 1950s, the German Marxist Ernst Bloch cheerfully foresaw a Promethean remaking of the earth through the use of atomic energy: Radkau, *The Age of Ecology*, 68–69.

[17] Judith Shapiro, *Mao's War Against Nature: Politics and the Environment in Revolutionary China* (Cambridge: Cambridge University Press, 2001).

[18] Jacob, *Histoire de l'écologie politique*. In earlier times, he wrote in summary, nature had been an issue exclusively for the right, but by the 1990s, it had become closely linked to some form of socialism (311, 312). Antoine Waechter, whom Jacob describes as politically a relatively conservative figure in the French green movement, nonetheless espoused an impeccably progressive program: Waechter, *Dessine-moi une planète*.

environmental engineering had long gone together.) Yet, by the end of the Reagan presidency, Drake observed, developments had made such a combination all but impossible and right-wing hostility to environmentalism had become almost monolithic.[19] Meanwhile, the long persistence in Russia of a regime at least nominally left-wing in ideology and Promethean in earthview has perhaps inhibited the developments seen elsewhere. As late as the 1980s, when the modern pattern was already apparent in the other countries examined, the mixture of sociocultural traditionalism and environmental advocacy seen in the Russian "village prose" school of literature—a mixture, as we have seen, not untypical of the European or American past, though unusual today—represented one of the only relatively safe channels for muted dissent against the Soviet system.[20]

The sociologist Robert K. Merton usefully warned that "establishing the phenomenon" should normally precede any attempt to explain it.[21] This book has not incontestably proven that pre-1960 Prometheanism was a predominantly left-wing earthview. But it has, at least, provisionally established it. That being so, the question that then arises deserves at least a few pages of discussion. Why is it that we see (if we do) such a different, indeed opposite, set of alignments prevalent today? Why have environmentalism and progressivism, Prometheanism and conservatism (if they have) swapped their partners from the past? Even if Dryzek is correct, and Prometheanism was ubiquitous before the mid-twentieth century, one must ask why it seemed consistent with progressivism then and does not today. The question is not one of merely historical interest, for an answer to it would help make sense of the present and, perhaps, the future. If, as this book suggests, we will not find some profound and fundamental reason for today's pattern of associations, something that makes Prometheanism inevitably a right-wing and environmentalism a left-wing earthview, then something more contingent must have played

[19] Drake, *Loving Nature, Fearing the State*. See also Turner, "The Specter of Environmentalism."

[20] Kathleen Parthé, *Russian Village Prose: The Radiant Past* (Princeton, NJ: Princeton University Press, 1992).

[21] Robert K. Merton, "Three Fragments from a Sociologist's Notebook: Establishing the Phenomenon, Specified Ignorance, and Strategic Research Materials," *Annual Review of Sociology* 12 (1987), 1–28.

a deciding role in the reversal that has taken place. If it changes again, further reversals may be possible. I do not have a fully worked out answer to offer, but rather some suggestions towards one.

One thing that has changed, of course, is the new extent and intensity of concerns about harmful human alterations of the environment: the rise, however unevenly, of an ecological consciousness that sees nature as intricate, complex, and not lightly to be meddled with. But such new beliefs cannot, by themselves, account even for the replacement of Prometheanism by environmentalism on the left, for whether or not they are valid is precisely one of the points in contention between right and left, between Prometheans and environmentalists, today. We must still ask why some people accept them while others do not. And even if this factor could explain what happened on the left, it cannot account for the concurrent rise of right-wing Prometheanism. If elements of conservatism as an ideology offer hospitable grounds for environmentalism—and they do, as the introductory chapter has shown—conservatives had more reason to hail the rise of a green mindset as a victory and a vindication instead of disassociating themselves from it.

The change in the USA and western Europe coincided, more or less, with the displacement of the Old Left by the New. A productivist, materialist, work-centered worldview highly compatible with Marx's (especially the later Marx's) environmental Prometheanism gave way to one often described as postmaterialist, more receptive to such new social movements as feminism, and much concerned with quality-of-life issues, among which environmentalism readily took a place. Jean Jacob's history of French political ecology has highlighted the importance of just this transition. It likewise had much to do with the realignment of the Democratic and Republican parties in the USA beginning in the 1960s, and the former's increasingly greater receptivity to environmental concerns.[22] Green causes also had much to offer the New Left's generalized affinity for protest and its particular aversion to the large corporations that could be blamed for many environmental problems. In the USA, some representatives of the Old Left, for their part, were able to maintain

[22] Jacob, *Histoire de l'écologie politique*; for the USA, see Adam Rome, "'Give Earth a Chance': The Environmental Movement and the Sixties," *Journal of American History* 90, #2 (2003), 525–554.

a characteristic earlier dislike of environmentalism intact as they metamorphosed into neoconservatives.[23] Correlation, to be sure, is not cause, and this correlation raises some questions as well as helping to answer others. The New Left was more deeply distrustful of government—as a part of "the system" it attacked—than environmentalists are, and liberation for the sake of personal gratification, hardly a theme congenial to environmentalists, was another pervasive element of its credo. And again, even if one could make sense of what happened on the left, what happened simultaneously on the right also demands explanation. If the left finally expelled Prometheanism, why did the right take it in instead of holding to its longstanding championship of limits, restraint, and the authority of nature?

Another change is that, with the tremendous growth in human numbers and technological capacity since 1800, Promethean dreams of having the kind of earth we would like no longer seem necessarily dependent on the resources, the credit, and the authority of the sovereign state. Significant environmental change, it is now clear, can be brought about by innumerable private actions coordinated only by the market, and indeed will occur if such actions are not restrained. The key question remains, though, whether the change these actions bring about is for the better or the worse. One of the basic contentions of modern environmentalism is that the market and suchlike hidden hands will not do the job of restraining all of the human actions that carry the danger of exceeding natural limits or provoking harmful side effects. At least some of them will need to be restrained by state action. But present-day market conservatives or Hayekian liberals, given their shared doubts about the ability of state planning to outperform spontaneously ordered action, would expect much less from it than a progressive might, either in restraining harmful impacts or in encouraging beneficial ones. They would more likely put their faith in the ability of societal feedback loops, those of Hayek's spontaneously evolved orders, to reward and encourage the good ones and at the same time modify and correct the bad ones. What has

[23] Eric Hoffer, "The Return of Nature," in *The Temper of Our Time* (New York: Harper & Row, 1967), 79–96 and "Cities and Nature," in *First Things, Last Things* (New York, NY: Harper & Row, 1971), 29–40; Richard Neuhaus, *In Defense of People: Ecology and the Seductions of Radicalism* (New York: Macmillan, 1971).

also changed since the period in question is that traditionalistic conservatives have abandoned the state, to which they once clung, and joined Hayekian liberals in their nearly automatic aversion to it, or at least to many of the things—notably taxation, economic regulation, and what one can call either the maintenance of social safety nets or economic redistribution—that it might do. (A strong suspicion of capitalism, conversely, was one of the elements common to the green conservatism of early nineteenth-century England, as well as to pre-twentieth-century conservatism more generally.) The more the state, when in the hands of the left, has undermined traditional arrangements and hierarchies, the more inclined traditionalists have been to throw in their lot with its classical liberal opponents, and the more reluctant they have been, as a result, to accept environmental regulation.

The most plausible explanation that I can see for the puzzle of Prometheanism's political associations lies here. For the reasons offered throughout the book, environmental Prometheanism, when all else is equal, finds an immediately congenial home on the political left, as it did in the past, just as environmentalism does on the right. But not all else is equal, one factor in the contemporary world in particular being salient and powerful enough to upend the expected pattern. Those who welcome and those who fear state intervention (or at least the kinds that people tend to think of as "state intervention")—in other words, the contemporary left and the right, broadly speaking—will not be equally inclined to accept an environmentalism that implies the need for a great deal of it.[24] The right will dismiss the environmentalist case, but, needing some coherent and defensible philosophy of nature and society to justify that dismissal, will adopt the most plausible one at hand, environmental Prometheanism.

The objection might be raised that the state, at the time most of the progressive Prometheans were writing, was still firmly in the hands of the small minority at the top, the rich and the well-born. Thus, one might expect to see them—as indeed such anarchists as Kropotkin and Reclus

[24] If the Polanyite critics are correct, "antistatist" remains a useful label, but not one that can be understood at face value, for those who adopt it identify themselves with a particular package of state actions while resisting others, the latter, in any case, including most of the kinds of government regulations that environmentalists urge.

did—distrusting it, still more any global government that might grow out of it, and reluctant to assign it the task of engineering the planet. But the radicals and progressives of the nineteenth and early twentieth centuries envisioned, among other things, the eventual capture of the state and its powers by the oppressed minority as part of the progress they expected. Only then did they suppose that it would bring about an environmental utopia on earth. Indeed, many explicitly tied the two together, seeing the present order as not only unjust, but unintelligent and incapable of successfully carrying out such a task. What might otherwise seem an anomaly, the support for population control by Wells, Gruening, and other progressive Prometheans, becomes understandable when viewed as another occasion for the introduction of rational planning into human society.

A tentative conclusion, then, would run something as follows. Promethean ways of thinking about the natural environment and progressive ones of thinking about society under normal conditions have strong affinities for each other, and so do environmentalism and conservatism. These associations, however, have been disrupted by the modern traditionalist right's embrace of the classical liberal market and its abandonment of the state—seemingly bent on uprooting the traditions that it most values—as a lost cause, making it virulently hostile to any beliefs that imply, as much of environmentalism does, that state action is needed to correct the damage that unregulated private action would inflict. For contemporary progressives, on the other hand, the state represents the best hope for achieving desired goals. They are as receptive as conservatives today are not to environmentalist calls for regulation, which offer a wider scope and a more pressing case for state action than does any version of Prometheanism currently available, and hence to the arguments and evidence that underpin them. Attitudes toward the state, as I argued in the first chapter, are an accidental, rather than a foundational, difference between left and right, but they are one to which the accidents of history have given an overriding importance in politics, and therefore in environmental matters, today.

This explanation supposes that most people's environmental attitudes are secondary and derivative, and their political allegiances primary and determining. Drake has discussed the politics of water fluoridation in the

USA, a particularly valuable test case because of the apparent incongruity of its ideological alignments with those on other issues of environmental risk.[25] American liberals have generally supported the addition of sodium fluoride to public drinking water, and protest has been most vocal on the right. Conservative opponents have invoked all of the arguments customarily employed by environmentalists against other suspected toxins, while liberals have not applied the precautionary principle here as they readily do elsewhere. The most plausible way to resolve the anomaly seems, to me, to be to focus on the factor I have relied on above: the contemporary attitudes on the two sides of the ideological spectrum toward the state. The right attacks fluoridation as a government program even while it acquiesces in exposure to toxins introduced into the common media of air and water by private enterprise. The left sees fluoridation as a model exercise of government power for the general good. In each case, consistent with my argument above, environmental and scientific considerations as such are largely trumped by those of political ideology.

Some readers will find the claim not merely plausible, but rather banal, particularly those familiar with the argument that the social arrangements we dwell under and the commitments we make do more to influence what and how we think—especially about matters with which we are acquainted only at second hand and on the authority of those whom we choose to trust—than what and how we think determines the ways we live. The intense frustration that scientists knowledgeable about the risks of human-induced climate change have felt at their inability to effect any meaningful response is surely a clue that there is something more to what people think and do than simple reasoning about the information that accredited experts transmit to the public. We not uncommonly speak and think, it is true, as if the realm of nature were some kind of bedrock reality compared with that of society. We more commonly act as if the opposite were the case and derive our picture of nature from our social relations.

It has become quite common for conservatives recently to argue in just these terms that environmentalism is a kind of Trojan horse, whose real purpose, hidden but steady, is to strengthen further the power of the

[25] Drake, *Loving Nature, Fearing the State*, Chap. 2.

state for the interests of those who hope to benefit from thereby.[26] It may seem that I am signing on to that thesis. But those who put it forward are—not explicitly, of course, and perhaps not even consciously, but by clear enough implication—advancing a second one as well that robs the first of its political force. It is that their own stance can be accounted for in the same terms, that their own anti-environmentalism stems from a similar source: in their case, a reflexive hostility to the claim that state intervention is necessary or desirable in any particular case. For surely they are not seriously asserting that right-wingers and left-wingers are essentially two different kinds of people, two different species really, the former alone worthy of the name *sapiens* and gifted with the ability to see things clearly as they are, the latter's view of everything colored and distorted by a monomania for an enlarged scope of state action on any pretext that might justify it. Or, if they are making such a claim, it is one that no one need take seriously once it is put into words. The question remains which biases in any particular case are more distorting. "Liberals may be as predisposed as conservatives to sift evidence through ideological filters," in Clive Hamilton's words, "but in the case of global warming it happens that the evidence overwhelmingly endorses the liberal beliefs that unrestrained capitalism is jeopardizing future well-being, that comprehensive government intervention is needed, and that the environment movement was right to raise the alarm about global warming."[27]

Thus, we return to a point raised in the introductory chapter. Environmental problems exist in a real world, though it is a real world to which we have access only through our minds and our ways of thinking. Hardly anyone has the authority of first-hand knowledge of any of the realms of nature-society interaction in question, and most people have none at all, so that the questions become ones of authority, and authority that tends to be granted on the basis of one's trust and allegiances in

[26] E.g.: "The economist, Walter Williams, accurately refers to environmentalists as 'watermelons'—green on the outside, red on the inside. For if they can convince sufficient voters that economic growth under capitalism entails baleful environmental degradation, they can thereby gain political/legal control over the 'offending' businesses and industries. Environmentalism, in short, represents a viable backdoor to socialism." Andrew Bernstein, *Capitalist Solutions: A Philosophy of American Moral Dilemmas* (New Brunswick, NJ: Transaction Publishers, 2012), 46.

[27] Hamilton, *Earthmasters*, 87.

other matters. Those who are otherwise inclined to fear and distrust certain forms of state regulation will be receptive to the claim that they are unnecessary. Such ways of thinking are powerful but not unalterable. If the book has proven nothing else, it is that being conservative and being environmentalist need not be a contradiction in terms, and that being a Promethean, for a conservative, entails keeping some rather alarming company from the past.

It hardly seems likely, though, that the aversion to "big government" of the modern right will change any time soon, or the distaste for environmentalism that follows along with it. Indeed, if any change is easy to imagine in the near future, it is on the left, once it faces up to the magnitude of what preventing climate change, say, would entail. Avowed political progressives are prominent among those now developing the discourse of "the good Anthropocene," the idea that humankind is now irrevocably a force shaping the planet's surface and must trade the task of preserving the environment as little altered as possible for that of deciding toward what human-determined ends to shape it.[28] If current developments continue, what we may see before long in mainstream politics is a divide, not between environmentalism (once again, in the sense in which I have defined it) and Prometheanism, the former having been abandoned by both sides, but one between two kinds of Prometheanism: on the right, the market Prometheanism with which we are already familiar, and on the left, a state Prometheanism that presents government direction and coordination as essential for achieving the goal of abundance and a high-quality environment shaped for human wants: something not much different, in short, from the past kinds of progressive Prometheanism that this book has chronicled.

[28] Thus John Dryzek, who in the second (2005) edition of his book, described environmental Prometheanism as exclusively a right-wing discourse, in the third (2013) edition has, while largely maintaining that position, also noted the recent emergence of a new variety that "would replace markets with states": Dryzek, *Politics of the Earth*, 60 (see also 58–59).

Index

W.B. Meyer, *The Progressive Environmental Prometheans*,
DOI 10.1007/978-3-319-29263-2

D

E